Hartwig Müller

Medical Gases

Related Titles

Mozzarelli, A.A. (ed.)

Chemistry and Biochemistry of Oxygen Therapeutics – From Transfusion to Artificial Blood

2011

Print ISBN: 978-0-470-68668-3; also available in electronic formats
ISBN: 978-1-119-97542-7

ISBN: 978-1-119-97620-2

Blumberg, L.M.

Temperature-Programmed Gas Chromatography

2010

Print ISBN: 978-3-527-32642-6; also available in electronic formats
ISBN: 978-3-527-63214-5

Hartwig Müller

Medical Gases

Production, Applications and Safety

Verlag GmbH & Co. KGaA

Author

Dr. Hartwig Müller
Gartenstr. 24
47506 Neukirchen-Vluyn
Germany

Cover
© iStock 37388066; sudok1

Library of Congress Card No.: applied for

British Library Cataloguing-in-Publication Data
A catalogue record for this book is available from the British Library.

Bibliographic information published by the Deutsche Nationalbibliothek
The Deutsche Nationalbibliothek lists this publication in the Deutsche Nationalbibliografie; detailed bibliographic data are available on the Internet at <http://dnb.d-nb.de>.

Print ISBN: 978-3-527-33390-5
ePDF ISBN: 978-3-527-67604-0
ePub ISBN: 978-3-527-67603-3
Mobi ISBN: 978-3-527-67602-6
oBook ISBN: 978-3-527-67601-9

Cover Design Adam Design, Weinheim, Germany
Typesetting SPi Global, Chennai, India
Printing and Binding Markono Print Media Pte Ltd, Singapore

Printed on acid-free paper

Contents

Preface

Medicinal gases are gases used in the medical environment. Their range is defined differently by the most important pharmacopoeias, the European, the American, and the Japanese Pharmacopoeia. The defined gases and the permitted concentrations of their main impurities differ slightly in most of the pharmacopoeias, as do the analytical methods.

This is quite surprising because medicinal gases are comparable simple molecules (or atoms, in the case of argon, helium, and xenon) and their manufacturing and purifying methods have been known for more than 100 years.

In spite of their widespread use, medicinal gases remained a part of specialist knowledge for over a century. Starting from the 1980s, major efforts were made in Europe, to tailor these gases with specific limits and properties, and this found a place in several national pharmacopoeias and the European Pharmacopoeia.

With the long experience of having worked with and worked in several expert groups for medicinal gases for more than 30 years, the author's aim is to summarize the different developments leading to the present status in the description of medicinal gases.

It is the intention of the author to introduce specialist knowledge of these gases to a new generation of pharmacists, engineers, and medical doctors. A fascinating class of substances, they require special handling to gain the full benefit of their unique properties.

The present book should help professionals keep things simple where required and to take special precautions where past experience has revealed risks, thus requiring a risk-oriented approach. The book is not to be taken as a set laws, but is meant for the purpose of guidance only.

August 2014

Hartwig Müller
Neukirchen-Vluyn

General Remarks

Medicinal Gases

Medicinal gases represent a small and sharply limited sector in our health-care routine. Inclusive of oxygen, medicinal gases, in spite of their fascinating properties, comprised a subject intended only for specialists, remaining widely unknown for a long time. This might lead users into difficult situations, when they are not properly informed about medicinal gases and their correct applications.

Oxygen is like Janus – two-headed in its properties: if its level in our atmosphere should drop even slightly, we would be in serious danger of losing consciousness and our lives, just because a small percentage of oxygen is lacking. On the other hand, should the level of oxygen exceed the usual level even by a small percentage in our environment, all combustion processes would become violently accelerated and impossible to stop by using standard fire-extinguishing tools. In addition, oxygen under pressure is capable of igniting a hellfire, combusting materials that are usually chemically stable; under pressurized oxygen, even metals could burn down like haystacks.

These examples show in fact our splendid adaptation to ambient conditions with 20.9 vol% of oxygen. Only slight changes may lead to serious consequences. If we now consider situations, were we are dependent on the "good" composition of our breathing air, for example, during diving or when connected to a wall outlet in a hospital, or even working at a place with minor contact to the outside, in a wine cellar or a cave, always when we are connected via the umbilical to a cylinder or to any other source of gas, then we may get an idea of how important it is to be conscious about the gases in the atmosphere that is surrounding us all the time, which is so important to sustain our lives.

Unfortunately, our senses are not in a position to help us in detecting most of the gases – only a few minutes breathing of oxygen-deficient atmosphere are enough to sustain a permanent brain damage. Should we suffer from sudden headaches, we would never believe that a possible cause could be oxygen deficiency or breathing of dangerous concentrations of a toxic and odorless gas.

In spite of all these factors, we think we can blindly trust any steel cylinder with technical oxygen or compressed gas. After having travelled a long and winding road of safety for medicinal products, we have now established detailed regulations and good routines to have the "good" gas available for use. Now it is up to us to use reliable sources, working with good practices and experts, to know the precautions to be followed.

We have seen people working with liquid nitrogen, not aware of the danger of asphyxiation and cold burn. Molecular cooking has been paid already with lives, due to bad handling of liquid nitrogen. Bad handling during transporting of gases in not well-ventilated vehicles can also result in the loss of lives. Often the amount of gas that develops from small quantities of cryogenic liquid is underestimated, or, in case of carbon dioxide, from cryogenic solid. Pressurized cylinders in medical use are usually small and thus often subject to being dropped. Should the valve get sheared off, a missile is created, causing a violent venting of the gas, propelling the cylinder through the air.

One of the aims of this book is to describe the most important techniques and strategies behind selling a safe gas cylinder and guiding on how to use it safely.

The objective of the book is not to go into the details of medicinal applications. Here are experts both in the pharmaceutical as well in the medical field, who already knowing the applications and adverse reactions. What we have seen in the past often was a lack of the technical knowledge to handle the gases and cryogenic liquids in the presence of a patient and of keeping all the regulations in mind which are relevant for the transport and use of pressure receptacles.

I hope this book will help to close a gap in the literature; the readers are invited to give their comments for further exhaustive treatment of this topic. Safety both for the medical professional and for the patient should be the accepted goal while using the benefits of medicinal gases.

I express my thanks to the publisher, who was quickly convinced by the concept behind the book, my wife, who was always ready to support my work with her enormous patience and charity, all my former colleagues who helped me to discuss the important subjects, and, last but not least, the industrial gas companies, willing to share their pictures and knowledge with me.

Hartwig Müller

Figure 1 Chemical Laboratory in the 16th century (1570 Johan v.d. Straet).

1 Medicinal Gases – Manufacturing

1.1 Where Do the Gases Come from?

Oxygen as the most prominent (medicinal) gas often lets us think that all other gases are of the same origin, coming right from the air. However, this is not the case. In addition to the very well-known oxygen, argon, and nitrogen, it might be advantageous to use other sources owing to technical or economic reasons. In addition, quite often specific methods of manufacturing or synthesis lead to specific types of impurities. In this chapter, we take a look at different gases, their sources, and the types of impurities that are characteristic of these different sources.

1.1.1 Gases Obtained from Air: Oxygen, Nitrogen, Argon, Xenon

Ambient air is a fascinating reservoir of numerous meteorological effects and has been known since the beginning of mankind. That atmospheric air is a gas and thus just a specific form of matter, and, moreover, that it not only contains water but is a mixture of different gases, was discovered only a few centuries ago.

Until the seventeenth century, it was a general opinion that air is an element, and as such indivisible. In laboratories as shown in the cover of the book (Figure 1), researchers like Jan Baptist van Helmont, (1577–1644) from the (then Spanish) Netherlands (Figure 1.1) recognized that gas is not a unique element but composed from different gases.

He noticed the difference between the chemical properties of hydrogen (developed by the reaction of hydrochloric acid and zinc) and carbon dioxide (developed by the fermentation of yeast) [1]. The two compounds had a physical property that was named "chaos" by van Helmont – a word that had the same pronunciation in Dutch as "gas," – and this became the term for this state of matter.

He discovered two gases, with almost similar physical properties as air, but with different chemical properties: hydrogen readily burned when ignited, while carbon dioxide remained chemically stable under most conditions, but giving a white precipitate with barium chloride solution.

Medical Gases: Production, Applications and Safety, First Edition. Hartwig Müller.

Figure 1.1 Jan Baptist van Helmont (1577–1644) [1].

Lavoisier *et al.* [2] discovered that air is composed of different gases in the late eighteenth century. They showed through chemical methods that there were at least two main components in the air, one being chemically reactive and the other one chemically inert [3]. Besides chemical absorption by specific reactions, air can be separated into its constituents by fractional distillation as with the liquids, to obtain its pure constituents, depending upon the knowledge of the art of fractionation.

As can be seen in Table 1.1, the major components of air have critical temperatures far below 0 °C. Above this temperature, no liquefaction of the gas is possible, indicating that the gas has to be cooled down first to below the critical temperature, before condensation starts, if the cooling is continued. Two well-known processes had been developed toward the end of the nineteenth and in the beginning twentieth century, respectively, by German (von Linde, 1895 [6]) and French (Claude, 1902 [7]) scientists.

While Linde's process works with a throttle to release the tension of the gas and to cool down the compressed gas (the Joule–Thomson effect), Claude's method uses an adiabatic expansion machine. The result is a "cryogenic" liquid with remarkable properties, having an average boiling point of about −194 to −185 °C. This liquid can be distilled in appropriate columns.

Table 1.1 Composition of ambient air, typical components [4].

Name of the gas	Content (approximately)	b.p. (°C)	b.p. (K)	
Nitrogen	78.1 vol%	−196	77.4	
Oxygen	20.8 vol%	−183	90.2	Σ 99.8 vol%
Argon	0.9 vol%	−186	87.3	
Carbon dioxide	390 ppm	−78	197.7	
Neon	18.2 ppm	−246	27.1	
Helium	5.2 ppm	−269	4.2	Σ 26 ppm
Methane	1.5 ppm	−162	111.6	
Krypton	1.14 ppm[a)]	−153	119.8 K	
Hydrogen	0.5 ppm	−253	20.4	
Carbon monoxide	0.2 ppm[a)]	−192	81.6	
Xenon	0.089 ppm	−108	165.1	Σ 1 ppm
Nitrous oxide	0.3 ppm	−88	184.7	
Sulfur dioxide	a)	−10	263.1	
Sulfur hexafluoride	a)	−63.9	209.2	
Carbon tetrafluoride	a)	−128	145.2	

a) Under influence of human activities: depending upon localization of sampling, carbon monoxide and sulfur dioxide can be detected near industrial activities in considerable levels under specific conditions, sulfur hexafluoride and carbon tetrafluoride are gases that often escape during aluminum electrolysis, while krypton is contaminated with the radioactive isotope Kr-85, emanated during numerous nuclear processes [5].

Some of the low- or high-boiling components are emitted during human activities (industrial activities such as coal mining (methane), aluminum electrolysis (carbon tetrafluoride and sulfur hexafluoride)) or general activities such as those of traffic, power stations, and incineration plants (carbonmonoxide, carbon dioxide, nitric oxides).

On a technical scale, the air separation plant is a huge setup with separation columns reaching heights often of 20–50 m. A typical construction is shown in the Figure 1.2. The sequential steps of the air separation process of the incoming air can be described as follows:

Those contaminants that would precipitate under cryogenic temperatures have to be removed first (water, carbon dioxide, sulfur dioxide, higher hydrocarbons). The low-boiling pollutants, such as nitric oxides, sulfur hexafluoride, carbon tetrafluoride, methane, and acetylene accumulate in the oxygen fraction of the column. As a consequence, all these impurities show very clearly the increase in air pollution.

While the main constituents of ambient air (except water) remain constant in their concentration (although they are part of huge biological and microbiological cycles, such as the nitrogen and oxygen cycles), trace impurities present often are dependent on the specific localization, which is, the positioning of the air separation unit (ASU).

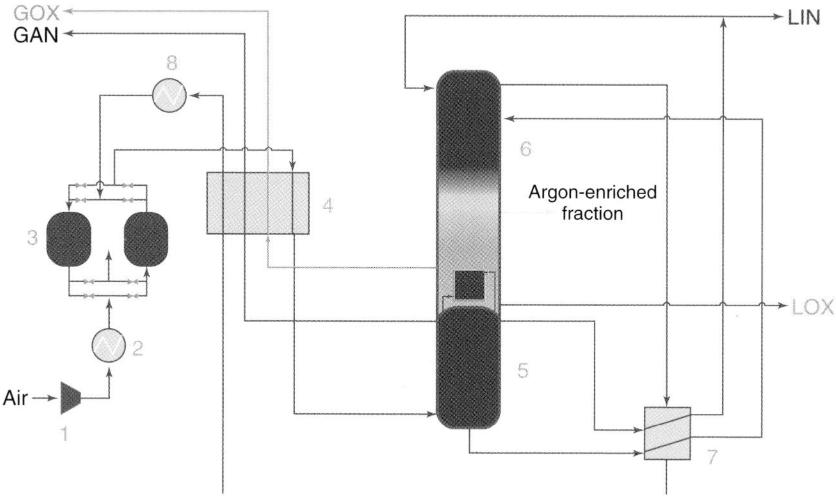

1 Compression
2–3 Purification. In the industrial environment the ambient air has first to be purified from dust, particles and high-boiling impurities, such as water, carbon dioxide, sulfur dioxide, and hydrocarbons, to prevent precipitation during the liquefaction process. Only the purified gas enters the liquefaction.
4 Liquefaction: Typically carried out at 5-6 bar (Linde-Fränkl process) to cool down the gas for liquefaction at about 100 K.
5 First stage of separation: 'high-pressure' column to separate oxygen from nitrogen. Nitrogen is sprayed into the top of the low pressure column to improve separation of pure oxygen in the
6 Second stage of separation: 'low-pressure' column to separate finally oxygen from nitrogen and to gain argon richh mixture.
7 Heat exchanger
Liquid Nitrogen (LIN)/Liquid Oxygen (LOX) Storage of the liquid product (oxygen, nitrogen)
The argon enriched mixture (GAN) is separated in an additional column, argon is purified by the reaction of hydrogen with oxygen and thus is pure argon created.

Figure 1.2 Air separation: schematic drawing [8].

Sometimes, these impurities undergo an increase or a decrease during a longer period of time, for example, since the 1950s, trace concentrations of carbon tetrafluoride and sulfur hexafluoride have shown continuous increase.

All nuclear events of the past released radioactive material into the atmosphere [8], krypton being the most prominent gas emanated. Radioactive isotopes of krypton kept changing in their concentration, depending on nuclear activities (atmospheric nuclear bomb tests in the 1950s, test ban in the 1960s, nuclear fallout following the Chernobyl disaster in the 1980s and the Fukushima blasts in the twenty-first century).

All these atmospheric ingredients, oxygen, nitrogen, and even the trace gases are integrated parts of appropriate cyclic equilibrium processes triggered by the dynamic chemical and meteorological phenomena of the atmosphere created by sunlight.

The fundamental starting point is that of cool gas sinking to the ground, while warm gas rises up in height; on the way upward, the warm gas is cooled down again and starts descending to the ground at another place. In addition, the possible content of water, as vapor or as droplets, which is directly linked to the temperature of the gas, is cause for many additional effects such as the generation of clouds and fog, as well as rain and snow.

Chemical effects are an integrated part of the meteorological effects in the air. Numerous reactive constituents forming reaction chains that are constantly stimulated by the UV part of sunlight, lightning in thunderstorms, and so on. All this depends on the height of the layer in the atmosphere.

Figure 1.3 shows the different layers schematically. Every layer has its own specific contribution to the chemistry of the atmosphere depending on the specific conditions (composition and physics, such as UV-radiation and water content). This leads to a gradient in the concentration of most trace gases at different heights above the ground.

The upper layers of the atmosphere being subject to high radiation (UV and shorter wavelengths) in some manner collect gases escaping from the lower layers, because of chemical inertness. Under the conditions prevalent in the stratosphere, these molecules react with other light gases such as ozone that collect there (Figure 1.3).

Owing to the emitted radiation, radioactive substances such as krypton-85 (Kr-85) can easily be traced down to a few atoms with appropriate counters. Kr-85 is one of the fragments created during nuclear fission. Since it is an inert gas, traces of Kr-85 are emitted at every nuclear activity. The radioactive half-life of Kr-85 being 10.7 years [11], nuclear disasters such as Chernobyl, Fukushima, and others lead to a temporary worldwide increase of the available Kr-85 in the atmosphere and thus to a change in the radioactive load of the krypton fraction in the ASU.

These global effects are more of academic than of practical interest. Krypton is isolated from the residual oxygen fraction in the ASU: even if Kr-85 can be detected owing to the high sensitivity of the measuring methods, the actual amounts remain below the threshold limit values in the gas phase. Oxygen generated in air separation plants all over the world is a unique product and contains no radioactive particles above the natural level. Owing to the physical properties of liquid air in the rectification column, the process of air fractionation is quite stable if the frame conditions are maintained such that other (chemical) contaminations are not present.

Any failure of compression, of pumps, or of valves leads to an instant interruption in the process and the column is transferred to a safe condition. Moreover,

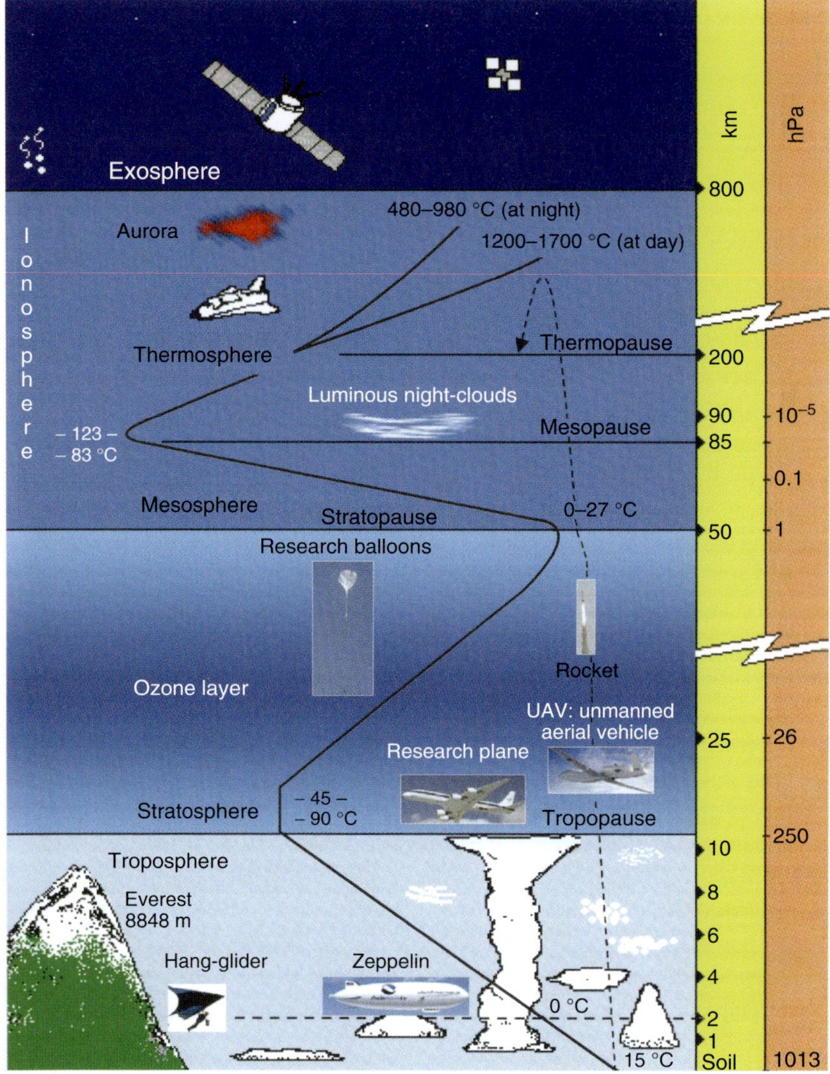

Figure 1.3 Different layers in the terrestrial atmosphere [10].

for reasons of safety, it is mandatory to check on-line the concentration of hydrocarbons in the oxygen (namely, acetylene that is collected) to avoid disastrous explosions.

1.1.1.1 **Oxygen**

Industrially generated oxygen contains argon as the main contaminant with the highest concentration of about 1 vol% maximum. The generated oxygen is suitable for big industrial needs of oxygen, namely, in the steel industry and, to some extent,

Figure 1.4 View of an air separation plant (Air Liquide).

the chemical industry, glass and ceramic industry, and finally the pulp and paper industry. The fraction of oxygen that remains for use as medicinal gas is thus only a small part of the overall oxygen consumption (Figure 1.4).

In spite of the importance of oxygen in medicine, there are only a few ASUs that work exclusively for the generation of medicinal oxygen. Medicinally used oxygen is just a by-product in the huge air separation process in most cases, which branches out beyond the plant, where it is then subject to further treatment following the GMP (good manufacturing practice)-rules.

Although consisting of the same molecules as industrial oxygen, medicinal oxygen is manufactured and processed as a drug, following the general GMP-Guidelines on the manufacturing of medicines for human use and also the specific GMP-Guidelines for medicinal gases, which will be explained in detail later in this book. In contrast to industrial oxygen, every batch of medicinal LOX and every cylinder of gaseous medicinal oxygen can be traced back to the generation and manufacturing process, because of an extended stepwise documentation of the manufacturing process including all checks and analysis made during generation.

Oxygen is marketed in Europe as a medicinal product both in liquid state as a cryogenic liquid or as a pressurized gas in steel cylinders (the pressure varying between 150, 200, and 300 bar). Although never administered to patients in its liquid form, the cryogenic liquid is also regarded in some countries as a medicinal product.

The cryogenic liquid itself requires some special care, as there are two main dangers inherent in this specific state of matter: the volume of gas generated even by small amounts of cryogenic liquid is quite considerable, one volume of LOX is vaporized forming a volume of gas that is almost 690-fold. This results in considerable oxygen enrichment, requiring a number of necessary precautions that are described later in this book.

The other main risk from a liquid at cryogenic temperature is the effect of the very low temperature on all sorts of materials, including human skin: on biological materials, a kind of cold burn is induced under sustained contact with the liquid (a few seconds are enough), while numerous materials show an effect called *cold embrittlement* when in contact with the very cold liquid and reaching cryogenic temperatures.

Owing to the low temperature, the material properties can be altered dramatically in a way that all mechanical strength is lost, and, for example, pipes lose their pressure resistance. All materials possibly or occasionally in contact have to be resistant to cryogenic temperatures, otherwise the access of cryogenic gas or liquid to such material parts has to be strictly prevented.

1.1.1.2 **Nitrogen**

The medicinal use of nitrogen is in most cases limited to application of its physical properties, namely, the transportation of cryogenic temperatures to any form of tissue to provide a sustainable conservation under low temperature, or, often under replacement of nitrogen by carbon dioxide of nitrous oxide, treatment of skin surfaces to clean up warts or similar irritations of the skin.

Although liquid nitrogen (LIN) is readily available, the handling of LIN, namely, the conservation of biological material requires sophisticated techniques to avoid destruction of the treated samples. Anyone who has already tried to freeze strawberries in a refrigerator would have experienced the difficulties of a proper freezing process, which is absolutely required so as to not to damage the cells by the sudden formation of ice crystals.

Special care has to be taken in the vicinity of storage containers to prevent hazards of LIN, especially in confined spaces.

The vaporizing liquid develops upon vaporization about the 850-fold volume of gas, creating a serious danger of asphyxiation, as the oxygen content of the air available for breathing will rapidly decrease in the vicinity of the container/s.

This risk is extremely high, when large amounts of substance are to be cooled down, thus warming up the liquid, generating large amounts of gaseous nitrogen, or during cooling down of "warm" containers, when all LIN in the containers has been removed by prior operations. Until the container has reached cryogenic temperatures again, LIN is vaporized in large quantities (Figure 1.5). The picture is showing on the left hand side a considerable flow of gazeous nitrogen leaving the receptacle, replacing oxygen in confined spaces.

Cryogenic temperatures on their own will create another risk, the danger of cold burns after contact with liquid or with cold parts of the piping or manifold. Cold burns are malicious, as because of an anesthetic effect of the cold, pain appears only after a while, when the burn has already taken place. It is a common practice to use the Leidenfrost phenomena for short contact with the hands or skin without any degradation of the skin: a thin layer of vapors isolates the exposed parts of the body, thus preventing for fractions of seconds deeper irritations such as burns or other such injuries.

Figure 1.5 Refilling LIN containers with cryogenic liquid (own picture).

In spite of a broad dissemination on the TV and Internet, this practice remains dangerous for two main reasons: bringing larger parts of the human body in contact with cryogenic liquid leads to accelerated vaporization of the cryogenic liquid, and the consequence is a highly possible oxygen deficiency of the environment, and last but not least, owing to contamination of the wetted skin, the generation of the isolating layer is not always the same, with possibly spots remaining with less or without isolation of the vapor layer, leading to severe burns of these parts.

1.1.1.3 **Argon**

Argon, being industrially the most widely used of the rare gases, finds use in a wide variety of applications, usually based on the ability of the gas to generate a stable electric discharge arc in an electric field. While this effect is used mainly for cutting and welding and in the lamp industry, it can be used also in the medicine under special conditions to close fissures in the tissue.

In clinical practice, this is done in surgery and at the dentist. The arc generates very high local temperatures leading to eschar at the edges of the cut or of the tissue.

1.1.1.4 **Xenon**

Xenon is present in air in orders of magnitude below the concentration of argon: it can be found only in trace concentrations, a generally accepted value of the content in ambient air is 0.08 ppm, meaning that a medium-sized presentation hall of 10 m length, 10 m width, and 5 m height containing roughly 500 m^3 of air contains about 40 ml xenon [12].

In specially equipped ASUs, a xenon-enriched liquid can be isolated, mainly consisting of krypton and oxygen. This liquid is often called the crude krypton.

Therefore, in a second additional process, this remaining enriched liquid is collected and processed again, to yield krypton and xenon in relevant quantities. This multistage enrichment process was originally developed for the isolation of krypton (see Figure 1.6, [13]). In a common layout, it first collects a fraction of crude krypton from one or more ASUs, which is liquefied again and then processed in liquid phase, separating the liquid into two gases: krypton and xenon. Xenon is kept back as a condensed liquid (Figure 1.6).

In the medicinal environment, xenon can be used in two different applications: for use as a contrasting medium in magnetic resonance tomography (MRT) (xenon affects the spin of the electrons). Xenon additionally has a market authorization in Europe as a mild anesthetic gas. In conjunction with other anesthetic agents, xenon gives a more tolerable anesthesia [13]. The narcotic effect can be supported with common liquid anesthetics to receive a very soft and smooth anesthetization, without some common, unwanted effects.

Technical applications not in the medical field include the filling of high-performance lamps (flashbulbs and headlights of cars) and extraterrestrial use as fuel in ion engines used for satellite movements in space.

Especially during anesthetization, large quantities of unwanted xenon would escape into the environment. As a consequence, an economic use of xenon in

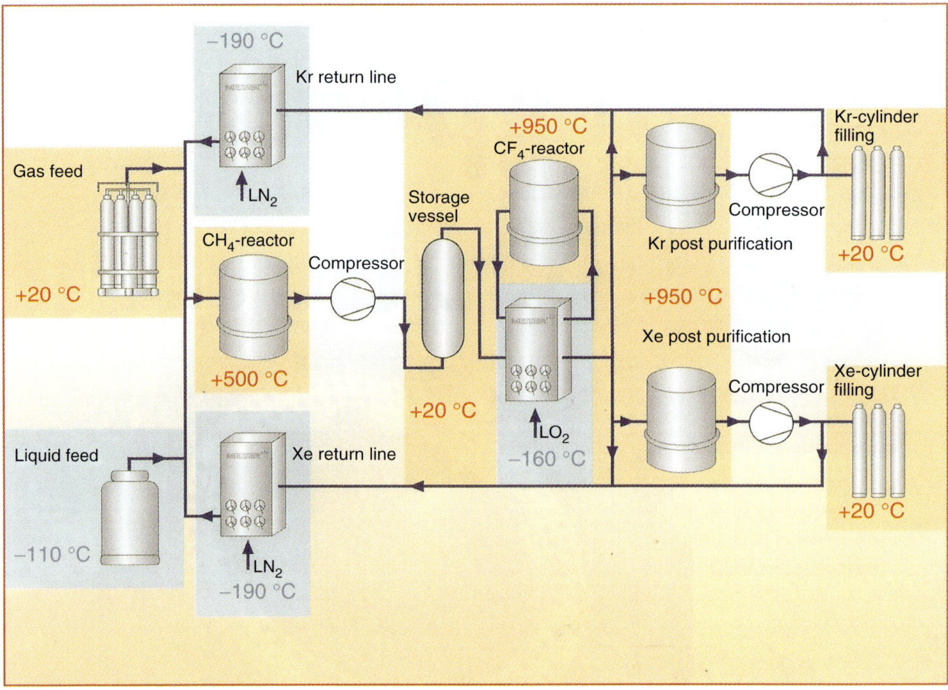

Figure 1.6 Separation of krypton and xenon [13].

Table 1.2 Physical properties of gases (I, [14]).

Name	Boiling point (°C)		Critical temperature (K)	Cylinder pressure (bar)	Status
Nitrogen	−196	77.4	126.26	200	Gas
Oxygen	−183	90.2	154.48	200	Gas
Argon	−186	87.3	150.86	200	Gas
Xenon	−108	165.1	289.74	52[a)]	Liq u pr

a) At 0 °C.

anesthetics is only possible with recirculation of nearly 100% of the used xenon in a closed-loop anesthetic device.

Xenon is marketed in cylinders liquefied under pressure. Some of the physical properties of xenon are listed in Table 1.2.

1.1.2 Gases Separated from Other Sources: Helium, Carbon Monoxide, Methane

Although most of the medicinal gases other than oxygen are present in traces in the ambient air, under the conditions of cryogenic air separation, the isolation, for example, of helium is not recommended: helium has a very low boiling point, thus it collects in the gas phase of the distillation column at a specific localization in the column. Larger quantities of helium (meaning here, a few liters) collected after appropriate periods of time form gas bubbles at the heat exchanger, thus decreasing the thermal conductivity and hence the efficiency of the unit. From this reason, helium (and hydrogen) has to be vented from time to time to eliminate the isolating gas, keeping the performance of the heat exchanger in the required range.

1.1.2.1 Helium

Helium was first identified on the sun, when, in the second half of the nineteenth century, Lockyer and Janssen discovered a new spectral line in the corona of the sun during an eclipse [14].

In the beginning of the twentieth century, it happened that helium came into the focus of military planning: lighter than air, it could be used to fill zeppelin-type airships and observation balloons. Helium is a rare gas and does not burn at all and so had been revealed to be a "safe" filling gas for airships at that time.

Military airships were not so common, because when filled with hydrogen, minimal sparks were sufficient to ignite the filling gas, creating a flash fire and causing the total burnout and loss of the ship. The disasters in the beginning of the twentieth century were legendary, but finally the crash of the "Hindenburg" during landing in Lakehurst, NJ, in 1937 brought an end to the filling of airships with hydrogen.

In our terrestrial environment, the amount of helium is limited. As no new sources of helium exist, it was decided that helium was a strategic war material. In consequence, a strategic reserve of helium was stocked. This "strategic" stock of helium in underground storages developed strong effects on the helium market in the world: by collecting or releasing amounts of the stored helium, the price development of the helium from third party sources [16] was strongly influenced.

With an average content of about 5 ppm in ambient air, helium is much easier to separate from natural gas, if this gas contains up to 0.1 vol% of helium, the method as such requiring about 0.04–0.3 vol% of helium. Isolating helium from gas wells with such low helium concentrations clearly needs a large quantity of natural gas to be processed, which is usually the case only in winter, when the consumption of natural gas rises.

Natural gas wells with a suitable concentration of helium and good productivity are quite well distributed all over the world, so helium needs very sophisticated logistics to be transported to the consumer. To save resources, helium is transported over long distances exclusively as a cryogenic liquid at the extraordinarily low temperature of approximately 4 K. This matches also quite well with the application of helium as cooling agent for superconducting magnets. Without the need for liquefaction on-site, helium only needs to be refilled as a liquid (Table 1.3).

At present, in large natural gas wells, when the main components of the gas (mainly methane and carbon dioxide) are liquefied, helium contained in the gas remains in the gas phase, is collected, purified, and also finally liquefied (Figure 1.7) [22].

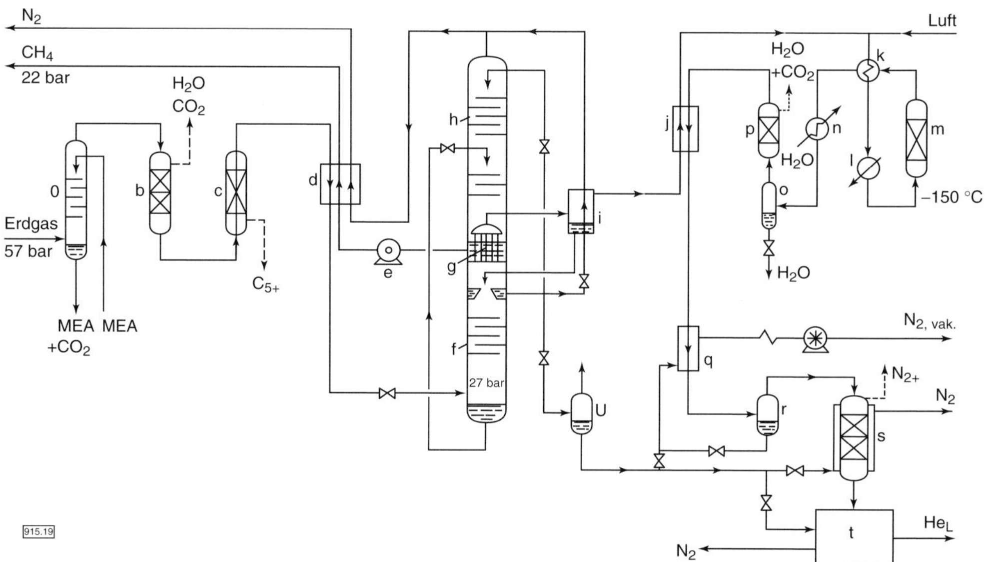

Figure 1.7 Separation of helium from natural gas [22].

Table 1.3 Exploitation of natural gas wells for helium [22].

Location	Liberal Kansas		Otis Kansas	Alfortville Frankreich	Ostrow Polen
Operator	n.a.		n.a.	n.a.	—
Manufacturer	n.a.		n.a.	n.a.	n.a.
On Stream since	1963		1966	1969	1975
Natural Gas	Kansas	Texas		Groningen	
Analyze (mol%)					
He	0.45	0.40	2	0.05	0.40
H_2	—	—	0.01	—	0.01
N_2	14.62	9.36	23	14.30	42.75
CH_4	72.74	79.26	68.5	81.15	56.01
C_2H_6	6.26	6.77	3.5	2.90	0.44
C_{3+}	5.73	4.01	2.5	0.70	0.99
CO_2/H_2S	0.20		0.5–1	0.90	0.30
Pressure (bar) abs.	45		20	41	57
Quantity ($m^3\,h^{-1}$)	950 000		25 000	65 000	137 000
Helium-Product					
Quantity ($m^3\,h^{-1}$)	6500		500	25	480
Purity (mol%)	65		>99 995	99 999	99 999
State of matter	gas		liquid	gas	liquid
Pressure (bar)	124		—	350	—

Figure 1.8 Helium tank trailer (own picture).

Helium is distributed in liquid phase in tank trailers (Figure 1.8), which are transported over sea by ship or by truck tractors over land. Large helium-containing natural gas wells are situated, for example, in the United States (Wyoming), in Russia (Sibiria), and in Qatar. The main consumers are concentrated in Western Europe, the United States, Russia, Canada, and Japan.

As mentioned above, helium is liquefied to optimize the transport. After liquefaction very pure helium gas is obtained when vaporized from the −269 °C cold liquid. The transport is managed with a kind of standardized tank trailers, each with a capacity of approximately 40000 l of liquid helium. Helium is transported in these tanks from the source to the customers everywhere in the world.

Helium is used as a refrigerant for superconducting magnets, as used in the medicinal MRT. To yield high resolution imaging, the magnetic field must be very strong. This is only possible using superconducting magnets. The solenoids are cooled down by liquid helium (temperature 4.2 K) to reach a superconducting state (Figure 1.9).

Before going onstream, the magnet is precooled with LIN, then nitrogen is removed by purging with helium, and finally, the cooling chamber is flooded with liquid helium. During normal operation, helium is gasified and evaporates in small quantities, depending on the efficiency of the isolation of the magnet.

An unwanted incident is the so-called "quenching" of the magnet. If the helium filling has fallen under a limit, parts of the solenoid can reach a temperature, where the superconducting property is lost. The resistance heats the solenoid very quickly and leads to a total warming of the solenoid, while more than 50% of the liquid helium filling is vaporized (as also the LIN shielding). The huge amount of gas generated must be safely vented.

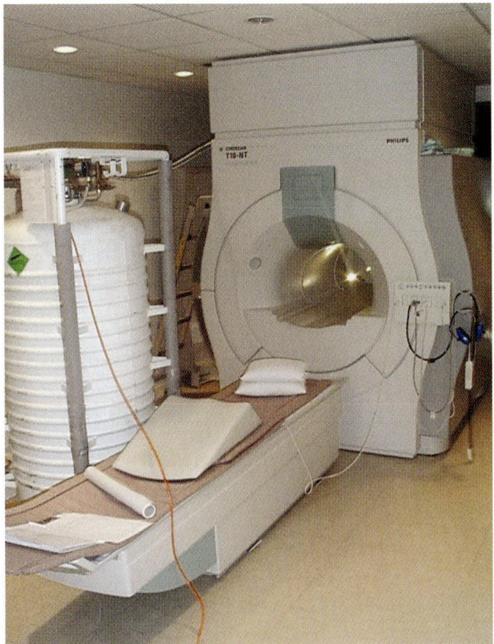

Figure 1.9 Superconducting MRT (magnetic resonance tomography) (15).

1.1.2.2 Carbon Monoxide

The old coke-oven processes disappeared in Europe parallel to the end of extended coal mining. Town gas has been substituted widely by natural gas, which is not as toxic and thus easier to handle. Carbon monoxide, therefore, nowadays is generated in the steam-reforming process, mainly used for the generation of hydrogen from natural gas. After removal of the sulfur components in natural gas (reduction to hydrogen sulfide and absorption with zinc oxide), the natural gas is mixed with hot water vapor and passed over a catalytic reactor in two stages to form a mixture of carbon monoxide and hydrogen.

According to the applied temperature, a mixture of hydrogen, carbon monoxide, carbon dioxide, and methane is generated, leaving the reactor at about 850–950 °C. The ratio H_2/CO can be varied according to the need of the application or hydrogen and carbon monoxide will be separated to yield pure hydrogen and pure carbon monoxide[1)] (Figure 1.10).

Carbon monoxide is purified from carbon dioxide by absorbers and then liquefied and rectified for high-purity carbon monoxide. The typical impurities in carbon monoxide are hydrogen, methane, and nitrogen under specific conditions (high pressure, steel cylinder) also iron-penta-carbonyl and nickel-tetra-carbonyl

1) Steam-reforming is a process used to generate hydrogen and carbon monoxide from natural gas or from higher hydrocarbons. The reformer contains a nickel catalyst, which transforms the steam–gas mixture into synthesis gas (H_2-CO·CO_2-CH_4), which is then processed again to yield either a mixture of hydrogen and CO or hydrogen and CO_2 (Air Liquide).

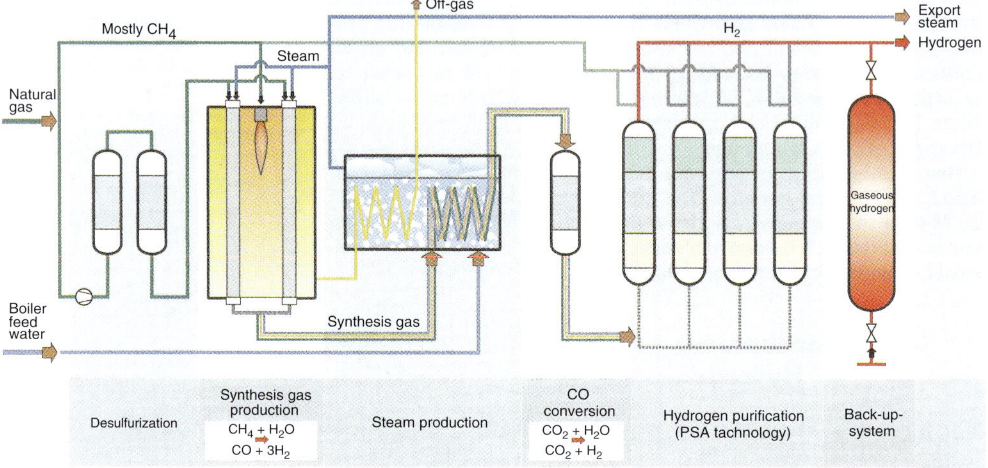

Figure 1.10 Schematic of the steam reforming process (1).

can be formed. Before medicinal use, all these compounds have to be completely removed.

A further purification can be done by liquefaction of carbon monoxide at cryogenic temperatures followed by a rectification. Cryogenic rectification is most often used to purify carbon monoxide beyond technical grades. Figure 1.11 shows a separation column from the outside.

Figure 1.11 Separation column for the purification of carbon monoxide [24].

1.1.2.3 Methane

Methane is a very common reaction product in various types of biodegradation. Natural gas, one of the most important sources of energy all over the world, contains up to 95% methane. It is liquefied and rectified to yield high-purity methane, suitable for all sorts of chemical syntheses.

Being a very stable molecule, methane can be found as a trace gas all over the world with an average concentration of about 1 ppm in the atmosphere. For industrial use, in most cases, pure methane is used. This is either generated by the reaction of carbon monoxide or carbon dioxide with hydrogen or separated from natural gas.

Natural gas consists of methane and carbon dioxide, in changing ratios, depending on its source. The composition of natural gas ranges widely from "dry gas," containing only methane and water in changing quantities, up to methane hydrates, "wet gas," accompanying oil wells and containing higher hydrocarbons up to C7 and carbon dioxide and sulfur compounds, and natural gas coming from condensate sources, containing hydrocarbons higher than C7 [24] (Table 1.4).

1.1.3 Gases from Chemical Synthesis: Carbon Dioxide, Nitric Oxide, Nitrous Oxide

1.1.3.1 Carbon Dioxide

Carbon dioxide is the stable residual of many exhaust gas purification processes in the chemical industry or a stable by-product of many chemical synthesis reactions. To yield higher purity, all contaminations (hydrocarbons, sulfur compounds, others) are oxidized in a catalytic converter under addition of oxygen and then separated (sulfur dioxide, nitric oxides, etc.) from the gas. From this reason a stable source with known composition should be used if the carbon dioxide is intended for medicinal use.

The fundamental steps for obtaining pure carbon dioxide are shown in Figure 1.12.

Table 1.4 Physical properties of gases (II) [14].

Name	Boiling point (°C)		Critical temperature (K)	Cylinder pressure (bar)	Status
Methane	−162	111.6	190.56	200	Gas
Carbon monoxide	−192	81.6	132.85	200[b)]	Gas
Helium	−269	4.2	5.20	200	Gas

a) At 0 °C.
b) Reduced for quality.

Although carbon dioxide is not toxic, breathing of higher concentrations than 8 vol% result very quickly in reactions within the body as carbon dioxide triggers a breathing reflex, and higher concentrations induce headaches, dizziness, and rising blood pressure.

Carbon dioxide is easily liquefied under pressure, filled in cylinders, and is marketed in steel cylinders of different volumes.

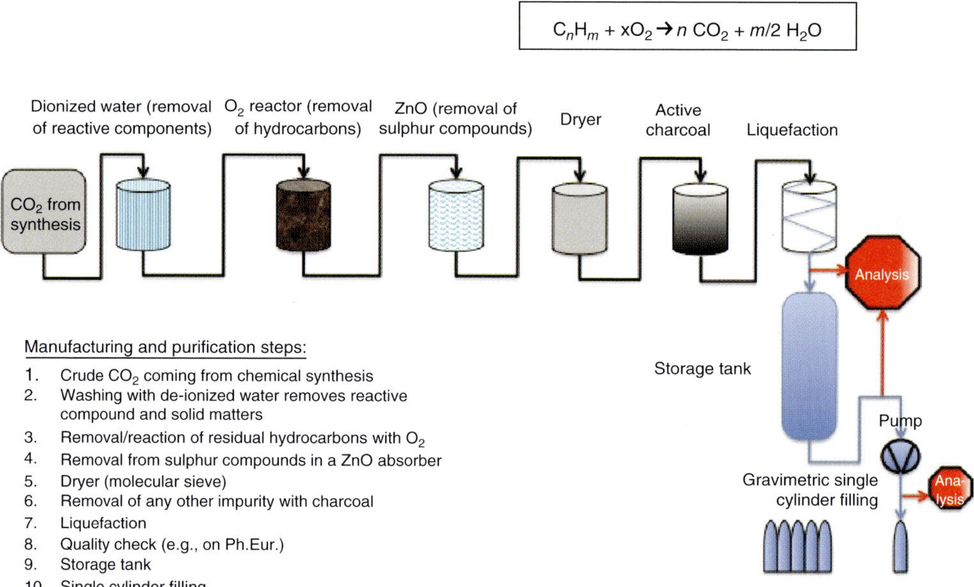

Figure 1.12 Purification of carbon dioxide.

1.1.3.2 **Nitric Oxide**

Nitric oxide is produced by the reaction of an aqueous solution of sodium nitrite and sulfuric acid or as an intermediate during combustion of ammonia for the synthesis of nitric acid (Figure 1.13). The generated gas is washed and dried in several stages and finally dried with a molecular sieve, compressed, and filled in steel cylinders. As nitric oxide is not stable under pressure, cylinders for highly pure nitric oxide are filled only to a pressure of approximately 50 bar, to keep the self-decomposition low, as this would contaminate the gas with other nitrogen oxides.

Nitric oxide must not be used undiluted. In lower concentration, nitric oxide can support specific breathing activities.

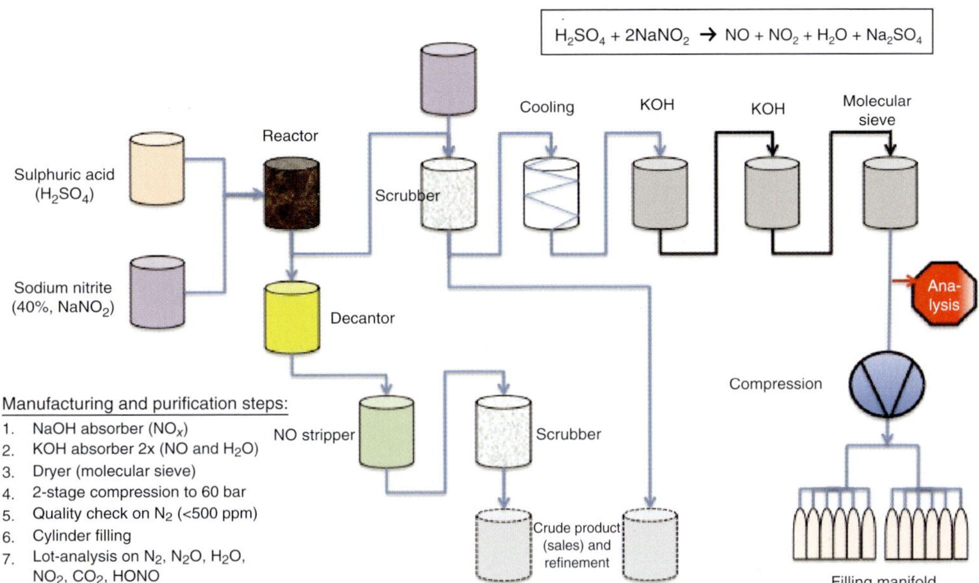

Figure 1.13 Synthesis of nitric oxide.

1.1.3.3 **Nitrous Oxide**

Although nitrous oxide is generated by numerous microbiological reactions in the soil, depending on the content of nitrogen and hydrocarbons [10] and thus released in the atmosphere, it is not separated from the air during fractional distillation.

Nitrous oxide is related to the fertilizer industry. Careful heating decomposes a concentrated solution of ammonia nitrate and the generated gas is nitrous oxide with trace impurities of nitrogen oxides and carbon dioxide (Figure 1.14).

Once these impurities are removed by washing, the gas is dried and liquefied and filled into appropriate steel cylinders as a gas liquefied under pressure. Nitrous oxide is stable under ambient conditions, but it supports fire and decomposes under high temperatures with tremendous release of energy.

The physical properties of nitrous oxide are similar to those of carbon dioxide, being a very efficient solvent in the supercritical state and liquefying/solidifying when the pressure decreases below a limiting value. The gas is liquefied under pressure and stored in cylinders (Table 1.5).

In the early days of gas therapy, it was Humphrey Davy, in about the year 1800, who first described the analgesic effect of nitrous oxide or laughing gas [28].

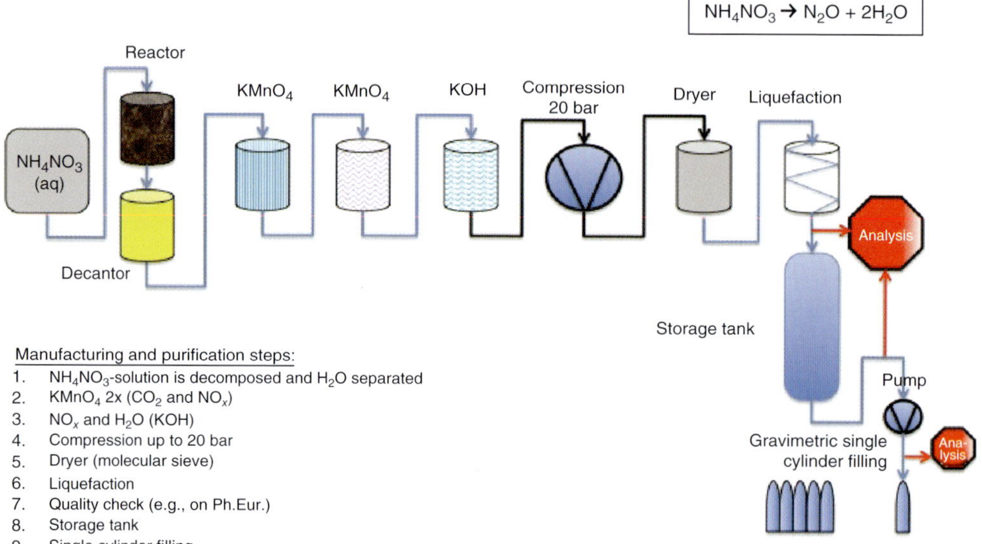

Figure 1.14 Synthesis of nitrous oxide (N_2O).

Table 1.5 Physical properties of gases (III) [14].

Name	Boiling point (°C)		Critical temperature (K)	Cylinder pressure (bar)	Status
Carbon dioxide	−78	197.7	304.21	55[a)]	Liq u pr
Nitric oxide	−152	121.4	180	200[b)]	Gas
Nitrous oxide	−88	184.7	309.56	55[a)]	Liq u pr

a) At 0 °C.
b) Reduced for stability.

1.1.4 Gas Mixtures for Inhalation

Except for oxygen, most of the described gases are not administered to the patient as pure gases, the reason being that any patient needs at least 20.9 vol% of oxygen in the gas he breathes in order to survive.

Any lack of oxygen is dangerous for the patient as all the organs, namely the brain, are very sensitive to oxygen deficit. Within a few moments, the level of oxygen in the blood is reduced to very low concentrations as the remaining oxygen in the blood vessels is further reduced by continued inhalation of gas without oxygen or with low oxygen levels.

So for reasons of safety, it is recommended that reliable breathing machines are used to mix gases at the operation theater or near the patient. In most of the European countries, gases for inhalation need a manufacturing license, as legally they are drugs because of their intended use for medicinal or diagnostic purposes so the breathing machines are medical devices acc. 93/42/EEC conso. 2007/47/EC [164]. Other gases are legally treated as medical devices, when they do not have a pharmaceutical effect on the patient, for example, carbon dioxide and some fluorohydrocarbons are used for dilatation purposes in diagnosis and micro-invasive surgery or for dilatation of the eye-ball often their application includes the appropriate medical device to facilitate the technical exercise.

Mixing gases is a matter for experts, as the gases have complicated physical properties, depending on their boiling, critical, and triple points. Wrong operation during the mixing procedure can lead to completely erroneous results, lack of preparation of cylinders can lead, for example, with reactive components, to instable mixtures, and, last but not least, mixing of oxygen and organic gases without thorough safety precautions can lead to serious explosions [29], often with drastic consequences.

In principle, the mixing of gases can be done either by weight ("gravimetric") or by volume ("volumetric") of the gases. Both methods having inherent difficulties in practice: while gravimetric methods suffer from the low weight of the gases (so $1\ m^3$ of nitrogen weighs only ≈ 1 kg) because of temperature-effects the volumetric methods are difficult to handle, because they need extraordinarily exact measurements, not only of the volume but also of pressure and temperature.

The fundamentals of how to prepare mixtures and perform measurements according to the Pharmacopoeia is laid down in the Annex 6 of the GMP and will be explained later. In fact, determination of the exact amount of the active ingredient is an important step during production of gas mixtures and puts high demands on the purity of the gases, the quality of the analytical methods used, in terms of stability and traceability of the calibration gases used, and in terms of the handling of the sample.

In other words, the use of ready-made gas mixtures, prepared and secured in their composition by reliable gas manufacturers, is recommended. This is true even for the simplest of all gas mixtures, recombined or synthetic air: in some European countries, recombined air is used as a substitute for compressed medical air.

For the substitution of mixing gases near the patient, ready-made mixtures for inhalation typically include the following:

1.1.4.1 Reconstituted (Synthetic) Air

It might be recommended to use recombined air when ambient air is polluted or loaded with contaminants, which is sometimes the case in summer in warmer regions. When mixing oxygen and nitrogen, the final mixture has to be analytically verified to be in the right concentration range of oxygen (between 20 and 22 vol% as defined in the appropriate Pharmacopoeia) to avoid any difficulty due to wrong oxygen concentration: an oxygen deficiency would create serious deterioration in the breathing of patients, an oxygen enrichment would create serious danger to the environment around the patient, as oxygen enrichment would cause a fulminate acceleration of fire if present.

1.1.4.2 Compressed Medical Air

Although a mixture, ambient air is usually regarded as a single gas, because during purification and compression, ambient air behaves like a single gas, by not demixing or decomposing during ordinary operations. The use of compressed medical air is only restricted by the quality of the ambient air, its contaminations, and water load. To remove these contaminations, information about the ambient air quality at the inlet of the compressor is needed, for example, if there are incinerator plants, heating installations, or laundries, in the neighborhood.

1.1.4.3 Nitrous Oxide 50 vol% in Oxygen

The mixture is used as a substitute to mixing gas with a breathing machine in cases where the application of breathing machines would be too complicated owing to the nature of the anesthesia required. This gas mixture is often used by practitioners or dentists. The use of the mixture is prevalent because it obviously breaks Dalton's law [30]: in a cylinder at room temperature, a unique and permanent gas phase with a pressure of up to 200 bar is available for use.

1.1.4.4 Nitric Oxide Approximately 1000 ppm in Nitrogen

This gas is used for inhalation to support dilatation of blood vessels in the lung environment. This gas must be used only with appropriate application devices to provide reliable mixtures with oxygen and to minimize formation of nitrogen dioxide by contaminated hoses or adapters.

1.1.4.5 Mixtures with the General Composition Carbon Monoxide, Helium in Synthetic Air (Carbon Monoxide Ranging between 0.2 and 0.3 vol%, Helium between 8 and 18 vol%)

These mixtures are used for diagnosis of lung functions. Depending on the type of instrument and the manufacturer of the device, they are used to measure the capacity of the patients' lungs by (blood) analysis during or after inhalation. Carbon monoxide is readily absorbed by erythrocytes, thus showing the equilibrium between offer of carbon monoxide and the value of occupied erythrocytes [25], equal to the absorption potential for oxygen.

1.1.4.6 Carbogen (5 vol% Carbon Dioxide in Oxygen)

In some mixtures, the positive effect of carbon dioxide on the breathing center is supported by adding small amounts of carbon dioxide to the gas (e.g., Carbogen is a mixture of 95 vol% oxygen and 5 vol% of carbon dioxide [26]). This concentration accelerates breathing which is in some cases of administering oxygen, a wanted effect [27]. Carbogen is used to facilitate exchange of gas molecules occupying erythrocytes. By the offer of a large excess of oxygen it is ensured that occupied erythrocytes are liberated by oxygen in very short time. As the exhaled gas contains only little carbon dioxide, this component has been added to support the breathing center, triggering deeper and more frequent breathing.

1.1.5 Gas Mixtures for Reference – Calibration Gas Mixtures

While all tests of the human and animal breathing functions rely on exact analysis of the composition of the exhaled gas in comparison to the inhaled ambient gas, reliable calibration gases are crucial to the determination of the concentrations.

As described above, Addition of carbon monoxide to the inhalation gas allows calculating the lung capacity via the calculation of the capacity to absorb carbon monoxide. Analytically, the ratio between inhalated and exhalated carbon monoxide is estimated and calculated.

Adding helium to the inhaled gas, the ratio of inhaled volume to lung volume can be calculated, thus the total active lung volume is estimated. Specialized analytical mass spectrometers are used to carry out all dilution and diffusion measurements simultaneously [31].

All electronically based analytical instrumentation needs gaseous standards to connect the reading of the instrument with the concentrations of the components. To accomplish this, all sorts of gases with carbon dioxide (often between 2 and 10 vol%) and oxygen (between 2 and 20 vol%) in nitrogen are now in clinical use.

The final determination of the ingredients of a gas mixture is as important as the determination of the concentration of the active ingredient. As a consequence, the manufacture of the gas mixture (preparing of the starting materials, mixing, and exact determination of the product's composition) needs thorough planning and exact handling.

The first step is to find the right material for the cylinder: this is needed to keep a stable composition of the gas mixture over the time. It is obvious that reactive components in the cylinder can react either with the cylinder wall or with other components in the mixture. This would alter the composition of the gas mixture over time, giving either incorrect analysis when used as a standard, or in case of therapeutic mixtures, one would be afraid of decreasing levels of, for example, the active ingredient. Often used materials for cylinders containing gas mixtures are either steel or aluminum, with specially treated surfaces that avoid chemical reactions with the components of the gas mixture.

To prepare gas mixtures, most often the components are prepared in a very pure state and then the mixture is prepared statically by adding one component after the other into the cylinders. For most of the gases with low boiling points, the composition is straightly depending on the partial pressures of the gases in the cylinder adding up to the total pressure.

For gases with higher boiling points, the so called "real gases," this simple equation has to be extended with correction factors to calculate the correct or required concentration. Namely, for carbon dioxide, nitrous oxide, and hydrocarbons, things turn out to be rather complicated and the mixture calculation is experts' knowledge.

Figure 1.15 shows the equations for a real gas, carbon dioxide, and the very obvious deviation from a linear plot as it is for "ideal" gases.

After the components are mixed and the balance added to the mixture, a thorough analysis is of utmost importance for a working standard. The calibration gas mixtures are used as calibration standards for any sort of analysis to establish a connection between the reading of the instrument and the concentration of one or more components in the mixture. This connection must be traceable to the SI System of weights and measures, to be robust enough to establish true analytical values of the subsequent analyses. These calibration gas mixtures thus are directly comparable to standard (volumetric) solutions used throughout the quantitative wet chemical analysis.

Figure 1.16 shows a typical linear calibration function with five calibration gases and one certified reference gas. Although the calibration gases are analytically connected to each other and to the reference gas, the effect is clear to see: the farther the value from the reference gas, the higher the analytical tolerance. Things

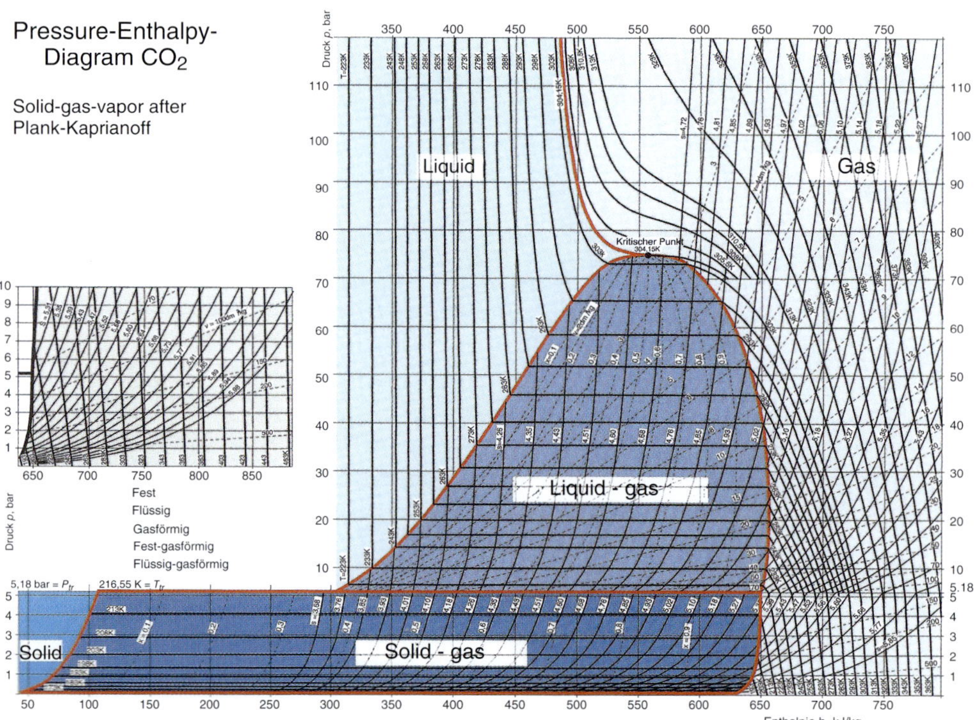

Figure 1.15 Equations of state for carbon dioxide [32].

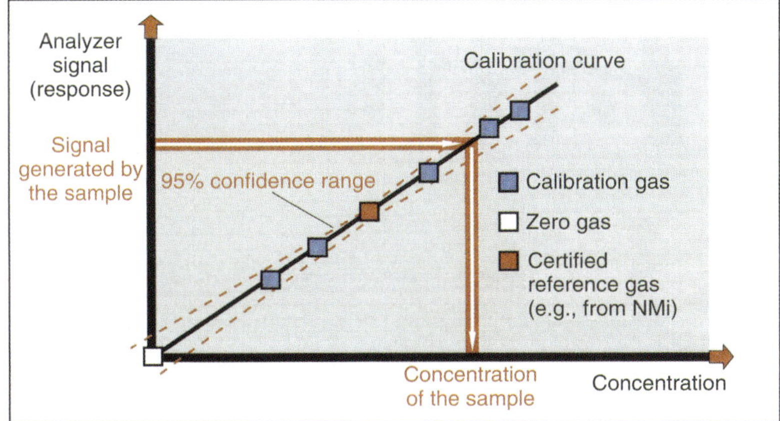

Figure 1.16 Calibration of instruments with calibration gases [33].

become more complicated, when the analyzer produces functions other than linear calibration functions.

For the measurement and good calibration of the instrument, the crucial factor is the withdrawal of the gas from the storage cylinder and the transfer to the instrument. All components of the mixture have to be transferred unaltered into the analytical instrument.

The use of a pressure-regulating device is mandatory from safety reasons: to avoid any damage to the instrument or even of the operators, there is no alternative to the use of a pressure-regulating device. Owing to their inner surface design, the instruments have to be purged several times by pressure buildup/venting cycles before the gas is led to the instrument for measurement [28].

2
Pressure Vessels and Their Accessories

In the middle of the nineteenth century, George Barth compressed nitrous oxide for storage under pressure in cylinders made out of copper [35]. From that time, gas cylinders were used for the safe storage and transport of gases.

This was the first step to separate the manufacturing of gases from their application. From this time on, there was no need for the doctor to generate gases near the application and the patient. At the same time, other difficulties arose: the transportation of the gas storage containers and the safe and reliable reduction of the pressure of the cylinder on-site.

Transportation, especially, remained cumbersome for a longer period of time: the weight and the properties of the cylinders were not common at the time, appropriate transportation systems had to be developed in the following years, and the weight of the cylinders had to be reduced.

To receive a satisfying ratio of the tare weight to the weight of the transported gas, the use of highly developed materials was necessary. Since the early days of the Taunton and Erhardt processes to make forged or welded steel cylinders, the weight of the cylinders has decreased considerably (Figure 2.1).

In those early days, thick-walled cylinders forged of steel might have reached a gas-to-packaging ratio of $0.05\,m^3\,kg^{-1}$, that is, that a cylinder weighing about 100 kg might have contained $5\,m^3$ of gas at atmospheric pressure [37]. Over the last 100 years, the industry has learned to use more sophisticated steels and to reduce the thickness of the cylinder wall, and, as a consequence, improved the gas-to-packaging ratio considerably.

The milestones in this development were the rise of the operating pressure from 50 to 300 bar and, finally, the introduction of the aluminum cylinder, the last one suffering from some material incompatibilities with reactive gases, but these gases are not relevant for the medical applications.

However, even today the unfavorable heavy package has not been replaced substantially by lightweight packaging. The lighter gases, helium and hydrogen, thus become a horror for number-orientated controllers: the amount of gas transported in a common 44-ton truck remains at about 100 kg of transported hydrogen, which is ridiculous compared to "ordinary" freight ratios, but is due to the very special properties of the gases.

Medical Gases: Production, Applications and Safety, First Edition. Hartwig Müller.

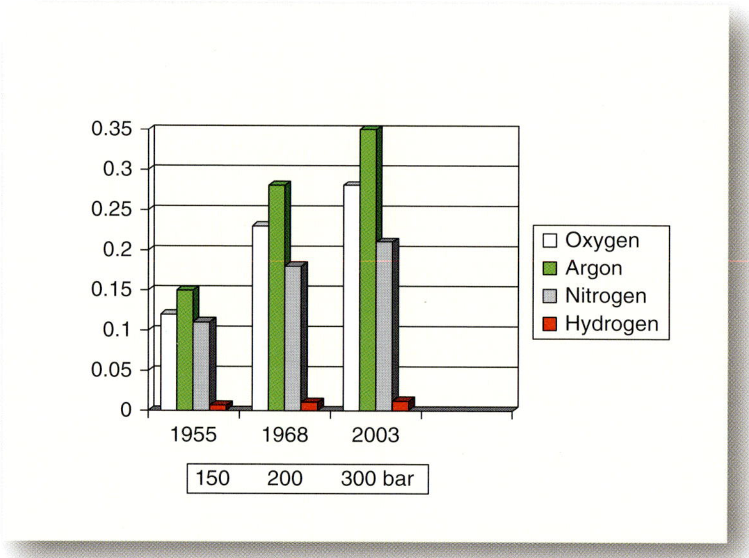

Figure 2.1 Ratio content to package – improvement by development of materials [36].

As one of the benefits of the work of the European Authorities, all the national regulations have been disappeared under European standardization by CEN/CENELEC. Since 1997 (stationary vessels) and 1999 (transportable vessels), all countries in the European Union and additionally those cooperating in the European Economic Region have the same standards validated (as always in the EU, this statement is valid for the great majority of rulings, but it cannot exclude some specific minor national exemptions).

The European Union has set up a series of guidelines to cover all sorts of equipment subject to pressure hazards, the TPED (Transportable Pressure Equipment Directive) (1999/36/EC, consolidated in 2011 to be 2010/35/EC [38]) and PED (1997/23/EC [39]), valid for transportable pressure equipment and stationary pressure vessels, was complemented by the directives for simple pressure vessels (2009/105/EC [40]) and aerosol dispensers (75/324/EEC [41]), respectively. Here the fundamental rules and construction characteristics can be localized.

In view of the already existing national legislations, the "new approach" to separate design regulations from operating regulations realized in these directives created a lot of changes to the approach of guidelines and how to set them into force. Especially in Germany, all inherited legislation such as the "Druckbehälterverordnung" [42] lost their validity, not so much because they became wrong, but just because of the other view of the legislator to create a flexible legislation that can easily be adapted by the member and associated countries of the European Union. All the old regulations and findings can still be referenced.

The core of the "new approach" is to avoid writing design solutions into legislation. On the contrary, a flexible regulatory environment should allow the European industry to develop new techniques and increase technical competitiveness. In fact, the EU Pressure Equipment Directives are part of a series of directives for the technical harmonization of machinery, electrical equipment, medical devices, simple pressure vessels, gas appliances, and so on [39]. All material built compliant to the European directives and standards carry at least the CE or an equivalent (Π-) marking (and other information, necessary for the user).

Except for material improvement allowing higher filling pressures with lower tare weights, there is no chance to improve the ratio for gases other than in small steps. In the past 100 years, only the switch to the liquid state of matter made things slightly better for the controller: cooling down the permanent gases to cryogenic liquids will result in 1 l of liquid nitrogen vaporizing into roughly 850 l of gas.

This switch of the state of matter for the gases liquefied under pressure is already inherent. Here, the economic storage size has been achieved under the pressure of the stored gas, so all advantages, mainly the high volume from a small amount of liquid, can be used [37].

To avoid difficult transportation of steel cylinders within hospitals, another kind of transport has become familiar during the last 100 years: the piping of gas from a central source to the point of use. Medical gas pipeline systems (MGPSs) have proved to be the ideal supplement to the supply of oxygen as a cryogenic liquid: cryogenic oxygen continuously vaporizes (ambient air has – even in winter – a sufficient temperature gap with the cryogenic liquid, so no additional warming is needed) and flows with constant pressure into the piping system of the hospital. This form of gas supply will be handled later, in a separate part of this chapter.

2.1
Transportable Pressure Receptacles: Pressure Cylinders

Behind the long-winded expression "transportable pressure receptacles" hide ordinary cylinders in steel or aluminum for transportation and storage of gases. The manufacture of pressure cylinders classically followed national standardization by the national standard institutes, as DIN (Deutsches Institut für Normung), BSI (British standard institute), AFNOR (Association francaise de normalisation), just to name a few, mainly for safety precautions to avoid the damage of cylinders during filling, transport, and use.

Presently, the world is on the way to a uniquely defined cylinder standard created by the United Nations [43], which will allow the use of cylinders (use of cylinders includes filling, emptying, testing, and refilling) at every place in the European Union and the rest of the world. Just to explain this statement: it is only a few years ago that refilling of cylinders coming, for example, from France to Germany, in Germany for the latter marketing in Germany was not permitted. French and German cylinders have been separated all along owing to secrets of manufacturing, which made the use of either of them in the other country unsafe for the authorities.

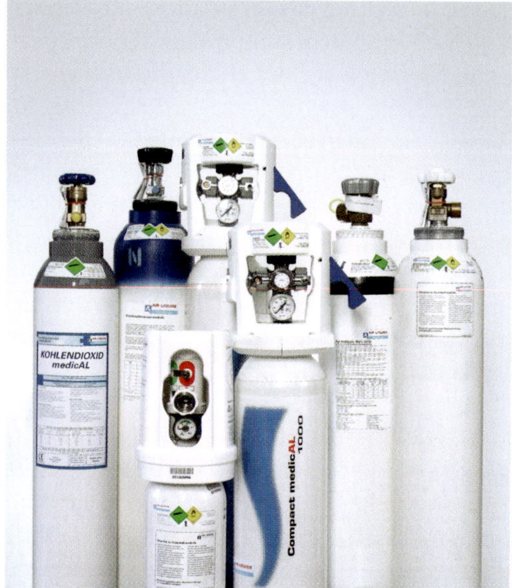

Figure 2.2 Medicinal gas cylinders (Air Liquide Medical).

From the view of the application in the hospitals, for oxygen, the smaller cylinders (up to 10 l water capacity) are used preferably in the vicinity of the patient, while big cylinders (50 l and bundles) are typically used in the central medical gas supply systems, where they are used to feed the piping network. Depending on the size of the hospitals, the feed-in with cylinders is often used only for an emergency operation, as a second or third source in accordance with EN ISO 7396-1.

In some countries in Europe, this fact has been introduced into the layout of the valve connections of the cylinders. While the small cylinders are equipped with pin index valves, the big cylinders are equipped with ordinary valves; in other countries, the small cylinders are preferably equipped with integrated valves, while the big cylinders carry ordinary valves (Figure 2.2).

2.1.1
Seamless Steel Cylinders

In the industrial gas field, the seamless steel cylinder is the core medium for transport and storage. In the last 100 years, this cylinder has reached a high level of safety when properly produced, that is, the required testing of the starting material and manufacturing processes are performed and all manufacturing steps are well organized and controlled.

Since the invention of the seamless cylinder by Mannesmann and others, the industry has operated this kind of cylinders in large numbers and considerable practical experience has been cumulated over this century.

Figure 2.3 summarizes the production process, as described in [37]. Faulty cylinders can be excluded during production as ultrasonic testing identifies them.

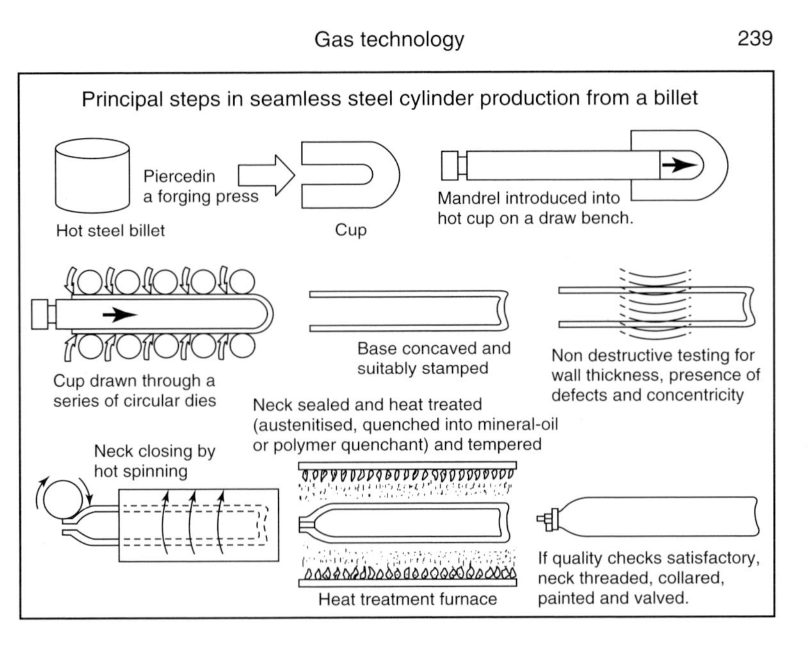

Figure 2.3 Manufacture of steel cylinders [37].

Minor quality issues show up in the forging of the bottom, where sometimes doubling or even holes can occur, when the production machinery is not thoroughly checked.

2.1.2 Seamless Aluminum Cylinders

Aluminum cylinders are produced from aluminum rods of specific diameter. These rods are cut into pieces and then formed into pre-material for the cylinders either by a hot or a warm process. Cylinders for medical oxygen service, 2 l geometrical volume, are exclusively manufactured in a cold process, similar to this one (Figure 2.4):

In the cold process, a press stamp is moved into the aluminum pellet, forming the bottom and the wall of the cylinder under a pressure of 2500 t.

To make this process smooth and easy, the stamp is wetted with special oils, which have to be removed prior to gas service for safety and for pharmaceutical reasons. In fact, a very thorough cleaning of the cylinder takes place.

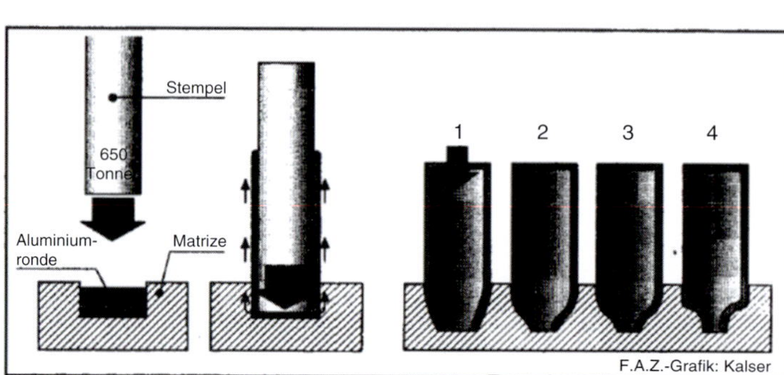

Figure 2.4 Schematic of Al cylinder process [45].

After the extrusion of the bottom, the aluminum body is cut to result in a clear segment and afterward, the upper part of the body is heated and another press forms the neck of the aluminum cylinder.

During the manufacturing process, appropriate cleaning is performed after several steps.

By using etching agents, such as sodium hydroxide or phosphoric acid, the inner surface of the container is cleaned and smoothened (Figure 2.5).

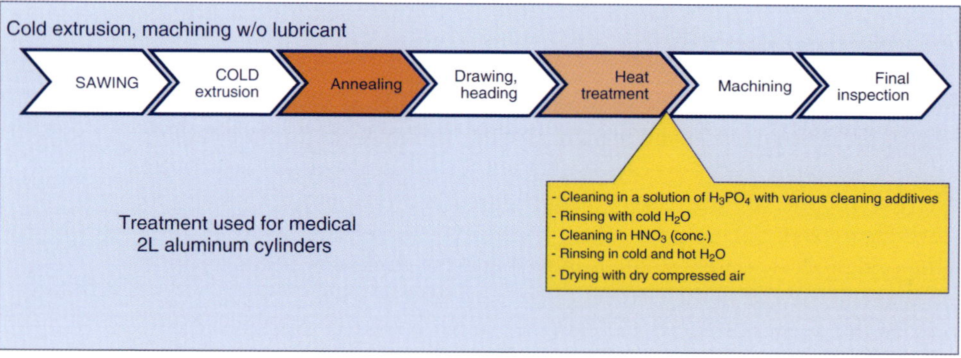

Figure 2.5 Cleaning methods for aluminum cylinders.

This is key to the use in oxygen service. Oxygen accelerates combustion of organic material when filling cylinders; high flow velocity together with material warming can easily result in a flash fire. By smoothening and washing away dust and debris, the number of particles in the cylinder is efficiently decreased. The micro particles play an important role in the generation of flash fires during withdrawal of oxygen.

The final stage of the manufacturing process is sintering in a furnace with subsequent cooling down in water, to give the aluminum alloy the necessary strength.

The key to lightweight cylinders is the used aluminum alloy. Compared to the AA6061, the new alloy AA7060 has superior material properties. With the use of special treatment, the strength of the material allows manufacturing lighter cylinders. For the user, this pays out in receiving of a lighter cylinder, which is essential, for example, for the home-care customers.

From the viewpoint of pharmaceutical use, cleanliness of the inner surface is essential. The several rinsing and purging steps during manufacturing have to completely remove all used oil and greases, which are necessary additives to perform the cold pressing of the cylinders.

2.1.2.1 **Specifics of Aluminum Cylinders**

Aluminum has different material properties than steel [44]. Besides the lower specific weight, a considerable difference in the reaction of the alloy toward heating or freezing can be noticed. While steel changes in its crystalline structure under influence of cold (usually −40 °C is regarded as the low temperature limit), aluminum is able to withstand even cryogenic temperatures (−196 °C) without loss of strength, but on the other hand, it is sensitive toward high temperatures. A temperature of more than 120 °C, applied for a day to a cylinder would essentially damage the crystalline structure.[1)]

Another point to be noted is that tensile strength of aluminum differs from that of steel. This effect is crucial when mounting valves. The valves have to be mounted with specific torque different from that used for steel cylinders, to avoid damaging the neck of aluminum cylinders, depending on the size of the valve thread (conical threads 25E or 17E). Care has to be taken also when applying the Teflon® tape to the valve thread to avoid any metal (brass/stainless steel) to metal (aluminum) contact. Moreover, stainless steel valves would damage the thread in the cylinder if an unsecured contact between the two occurs.

Aluminum cylinders are not forged; they are extruded with a hydraulically operated press resulting in a higher surface quality than that which is reached for steel cylinders. This very smooth surface can easily be purged, rinsed, and cleaned and is thus a good prerequisite for gas cylinders.

Aluminum cylinders are in widespread use in medical oxygen services and well-known in literature. Owing to their material properties, aluminum cylinders have

1) Following the recommendation of the manufacturers, aluminum cylinders should not be heated excessively; long-term heating should remain below 100 °C, so that no changes in the crystallization lattice can occur. In contrast to steel cylinders, aluminum cylinders do not undergo changes in tensility or strength down to temperatures until −196 °C.

Table 2.1 Commonly used aluminum alloys (AA wxyz, acc. EN 1975:1999 + A1:2003).

		Chemical composition (mass%)													
AA		Si	Fe	Cu	Mn	Mg	Cr	Ni	Zn	Ti	Zr	Pb	others		Al
													Each	Total	
6061	min	0.4		0.15		0.8	0.04								B.
	max	0.8	0.7	0.4	0.15	1.2	0.35		0.25	0.15		0.0030	0.05	0.15	
7060	min			1.8		1.3	0.15		6.1						B.
	max	0.15	0.2	2.6	0.2	2.1	0.25		7.5	0.05	0.05	0.0030	0.05	0.15	

a very smooth inner and outer surface, which is well protected against any attack of oxygen or moisture by a thin but very resistive aluminum–oxide layer of a few molecules diameter. This surface coating protects the surface from attack by the gas and also the transport of impurities from the cylinder wall into the gas phase (Table 2.1).

2.1.3
Welded Steel Vessels

The welded gas vessel is the third type of container that is often used to store and transport liquefied gases under low pressure. The welding seams clearly indicate the different parts of the cylinder: shoulder, mantle, and foot. Typically, high-pressure gases and permanent gases are not very often stored in welded cylinders (exemptions are the welded stainless steel cylinders for ultrapure or chemically reactive gases or large-size storage tanks for gases such as hydrogen). Welded pressure vessels are also used for the storage of cryogenic liquids. In this case, the inner vessel is surrounded by an outer vessel; the space between them is filled with isolating materials, such as perlite, and it is evacuated to prevent any transport of energy form the outside to the inside (see Figure 2.6).

Cylinders with a filling mass, usually in service for acetylene, are not used in medical environments. The filling mass is used to extend the surface of the solvent (acetone) used to dissolve acetylene.

2.1.4
Lightweight Wrapped Steel or Aluminum Cylinders

During the last 20 years, a new type of cylinder has been marketed: the wrapped cylinder. These cylinders consist of a metal liner to form a containment holding the gas; the wall thickness is laid out in a way that an additional reinforcement is needed to make the containment stable against the pressure of the gas. The resistance against the pressure is achieved by wrapping the cylinder with appropriate threads of glass or carbon fibers embedded in synthetic resins that harden out after wrapping.

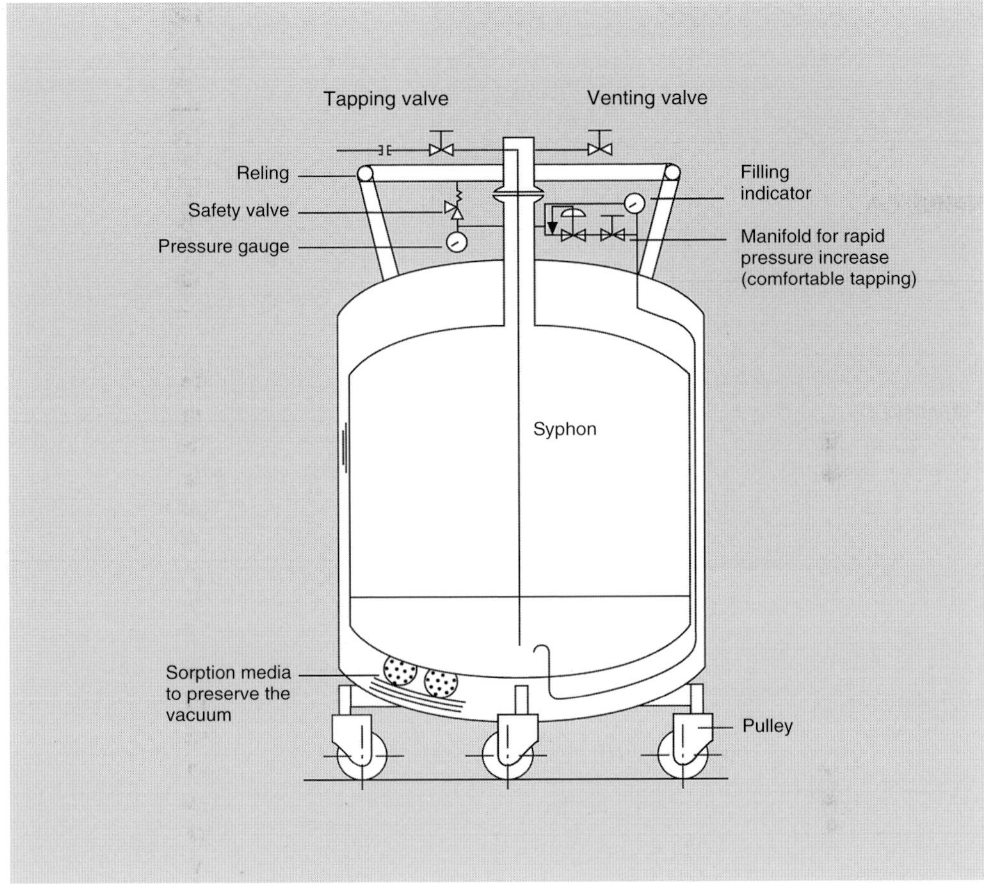

Figure 2.6 Small cryo-container [14].

Two main ways of wrapping have been developed, the fully wrapped container and the hoop-wrapped container, the liners varying between steel and aluminum (Figure 2.7).

Clearly visible in the figure are the two different materials of the liner (metal, on top – the stamping of the cylinder is on the shoulder, while the attribute of the wrapping is the inlaid binding with the writing).

While the low weight and high-pressure resistance of this kind of containers are well acknowledged, their sensitivity to mechanical shocks, cuts, and fire is a disadvantage in daily (clinical) life. Cylinders are often subjects to shocks and in the patients' environment, unsecured cylinders may fall down. In some countries in the European Union, the use of wrapped cylinders is restricted in certain parameters such as the time for retesting and the visibility of external shock marks.

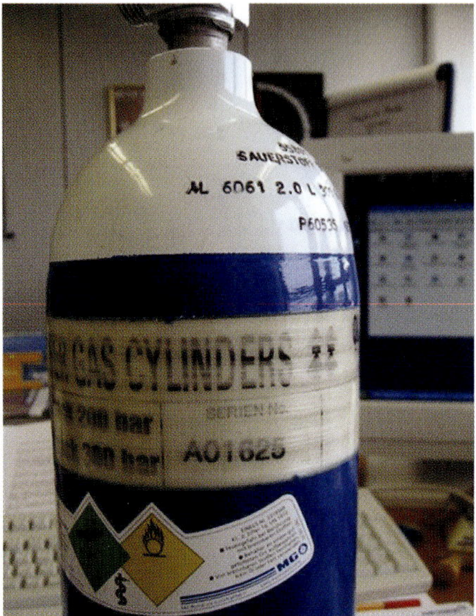

Figure 2.7 Wrapped cylinder (own picture).

2.1.5 Pharmaceutical View on Cylinders as Containment for Drugs

Pressurized cylinders are subject to numerous European and national regulations, concerning their transport (ADR), their storage, and their handling, with the use of words that are not very common in a medical environment.

All the technical standards that have been laid out under the TPED [38] are not so focused on the medical use of gases. There are some problems arising from the more technical nature of transport, which are solved differently in the European countries and which cause problems when cylinders are used in environments close to the patient without further precautions for keeping them safe and clean during transport, moreover, when taking into account that these cylinders are reused because of their high quality materials.

Additionally, medicinal gas cylinders have to comply with certain quality requirements derived from the regulation for the production of medicinal products (65/65/EWG [46], now 2001/83/EU [47]), and the GMP (good manufacturing practice)-Guideline.[2)]

The container for medicinal gases is somewhat special from the viewpoint of pharmaceutical literature: obviously, the cylinders are uncomfortably heavy and their handling is bulky, from reception to storage application and returning the empty cylinders.

2) Eudralex is the collection of pharmaceutical legislations in the European Union, governing medicinal products for human and veterinary use. It consists of 10 volumes.

From the viewpoint of the hygienist in a hospital, it seems reasonable to organize the reception, storage, release, and return in a way that minimum hygiene precautions must be respected: separation of cylinders in areas without possible contact with patients, for the use in the vicinity of patients, and for the use in critical areas (typically, where there is a presence of pathogenic germs).

Industrially manufactured gases in most cases do not experience any special treatment of the outer cylinder shell in view of possible germ contamination. If there is special care needed, the user of these gases has to set up a catalog of procedures before bringing these cylinders into critical areas, and, before returning cylinders from critical areas back to the supplier's storage. As the supplier, when handling cylinders returned from a hospital, does not usually take any special precautions, the user has to take care that all returned cylinders are impeccable from the viewpoint of contamination with germs.

2.1.5.1 European Pharmacopoeia View on Cylinders as Containment for Drugs

Despite these inherent questions, the European Pharmacopoeia just contains only general statements in Chapter 3, concerning packages made from plastic materials or made from glass [48]. Reusable steel or aluminum pressure vessels are not subject of the specifications.

Compressed gas cylinders are typically subject to transport operations. That is the reason ADR[3)] is the leading regulation for design and safety of pressure vessels and for amending the rules of the European Directive TPED. Besides the treaty wording itself, ADR consists of two main parts, Annexes A (defining the involved merchandises, their labels, and packages) and B (regulating the design, equipment, and use of the vehicles for transport of dangerous goods) and a further nine appendices in which the framework is defined, starting with terminology and classification of dangerous goods.

Gases are assigned to Class 2, "Gases," subdivided into flammable, non-flammable, and toxic gases.

ADR is the main reference for all transport-related questions. The type of labeling, which is dependent on the classification prescribed in detail, form, and size of the placards is defined in Annex A.

2.1.5.2 Inner and Outer Surfaces of Cylinders

From the point of view of the gas expert, it might be advantageous to make a difference between preparation and adequacy of the inner and the outer surfaces of cylinders for pharmaceutical use.

While the inner surface is prepared once during the lifetime of a cylinder (here, the lifetime is defined as the time between two periodical tests of a cylinder; this is a test performed from safety reasons usually for most of the gases every 10 years:

3) ADR: the European Agreement concerning the International Carriage of Dangerous Goods by Road (Accord européen relatif au transport international des marchandises Dangereuses par Route), ADR, is revised every 3 years; at present, ADR 2011 is in force.

Figure 2.8 Cylinder contaminated with blood.

the valve is removed and the cylinder is flooded with water to perform the pressure test; it is then dried and internally inspected, sandblasted in case of internal contamination with rust, and, once the cylinder is equipped with a new valve, the cylinder is purged with service gas and returned to service), the outer surface is cleaned or painted several times depending on the conditions of use and on how stable the initial painting was.

During use in the hospital, the outer surface of cylinders is subject to numerous opportunities of contamination, especially in the case of small cylinders up to 10 l: in the worst case of a cylinder ending up on or even under (to prevent it rolling away) the blanket of the patient. Treatment of the outer surface should respect this practice to avoid any displacement of germs both from the inside a hospital to outside it and vice versa. Figure 2.8 shows a heavily contaminated cylinder received from a hospital.

A rigorous training of all involved people should also help in the use of the accessories of the cylinder according their destination; a cylinder must not touch neither the bed of the patient nor the patient.

Except for occasional contact of the inner cylinder surface with ambient air or water during testing, reactions with the pressurized gas could occur. The numerous checks with oxygen seem to confirm that for reaction with oxygen, contamination with water has to occur first. Without water, no intermediate reactions with oxygen could be observed. In consequence, all efforts of the gas manufacturer are focused on preventing water from entering the cylinder by chance or by fault.

Reactive components in gas mixtures especially might also react with the cylinder walls. In most of these cases, there would be no harm to the patient, but more often, the reacting component could partially or completely disappear from the constituents of the mixture, leading to erroneous results of wrong calibrations. This instability of gas mixtures is usually avoided by using certified mixtures with a stability statement by the manufacturer; in most cases, these reactions are safely restrained by the use of aluminum cylinders which have a very smooth and chemical-resistant (i.e., resistance to most of the gases) surface.

2.1.6
Accessories for Cylinders: Valves

The most important accessory is the valve on top of the cylinder. As with cylinders, valves also went through development over the years toward safety of use and improved tightness until the multifunctional types with integrated pressure regulators that are produced now. We take a look at the different types of valves used in the medical field and their characteristics. Table 2.2 is an excerpt from the bigger standard systems used in France, Germany, the United Kingdom, and some equivalent ones, for different gases.

Things are more than complicated because in contrast to the streamlined regulations concerning containers (TPED, PED), valve connections in different countries have still not been standardized. In the beginning of the twentieth century, a variety of six or seven different valve threads was used to create the three major standard systems (AFNOR (Association francaise de normalisation), BSI (British standards institute), DIN (Deutsches Institut für Normung); among other associated ones such as for example in the Netherlands (NEN), Belgium (NBN), Sweden (SIS), and Italy (UNI). Originally used for a safe distinction among the gases, these different connections lacked international standardization and led to myriads of incompatibilities for one gas.

On the other hand, the use of integrated cylinder valves would replace all different connections (standardized in the different countries only for working pressure up to 200 bar) by low-pressure connections, which are now available at the cylinder.

That neither the industry nor the European legislator succeeded in creation of a unique connection for the most important medical gas, oxygen, clearly highlights the disarray in this area. Moreover, in the United Kingdom and the Commonwealth, a completely different valve connection system has been widely introduced for breathing gases – the pin index type of valve (Figure 2.9):

The dark points in the drawing represent holes, in which the rods of the counterpart are rested, when the valve is fixed (via a clamp). If the rods do not match with the holes, a wrong combination has been chosen and a connection is impossible.

The most important requirement (besides tightness and safety from flash fire) is a simple, good, and definite distinction between the different gases: no intermixing of gases must occur. This is provided by individually shaped pins on the regulator side (depending on the gas used), which must fit to the holes on the valve on the

Table 2.2 European Valve Connections (a, Continued).

Gas	FR/GR/PT	BE/LX	ES	IT
Air	C(nf)/D(nf) IS 21.7 × 1.814 RH fem	B4 (nb)/C6 (nb) IS 21.7 × 1.814 RH fem IS 30 × 1,75 RH fem	B(ne) IS 30 × 1.75 RH fem	UNI 4410 W30 × 1.814 RH fem
Argon	C (inf) IS 21.7 × 1.814 RH fem	B4 (nb)/C (nf) IS 21.7 × 1.814 RH fem IS 21.7 × 1.814 RH fem	C(nf)/DIN 6 IS 21.7 × 1.814 RH fem W21.8 × 1.814 RH fem	UNI 4412 W24.51 × 1.814 RH fem
Carbon Dioxide	C(nf)	B5(nb)/DIN 6	C(nf)	UNI4406
	IS 21.7 × 1.814 RH fem	W21.8 × 1.814RH fem W21.8 × 1.814 RH fem	IS 21.7 × 1.814 RH fem	W21.7 × 1.814 RH fem
Helium	C(nf) IS 21.7 × 1.814 RH fem	B4(nb)/C(nf) IS 21.7 × 1.814 RH fem IS 21.7 × 1.814 RH fem	C(nf) IS 21.7 × 1.814 RH fem	UNI4412 W24.51 × 1.814 RH fem
Nitrogen	C(nf) IS 21.7 × 1.814 RH fem	B4(Nb)/C(nf) IS 21.7 × 1.814 RH fem IS 21.7 × 1.814 RH fem	C(nf) IS 21.7 × 1.814 RH fem	UNI4409 W21.7 × 1.814 RH male
Nitrous Oxide	G(nf)	B4(nb)/G(nf)/C(nf)	U (ne)	UNI9097
	IS 26 × 1.5 RH male	IS 21.7 × 1.814 RH fem IS 26 × 1.5 RH male IS 21.7 × 1.814 RH fem	W 16.66 × 1.337 RH fem	W 16.66 × 1.337 RH fem
Oxygen	F(nf) IS22.91 × 1.814 RH male	A1(nb) IS22.91 × 1.814 RH male	F(nf) IS22.91 × 1.814 RH male	UNI4406 W21.7 × 1.814 RH fem
Xenon	C(nf) IS 21.7 × 1.814 RH fem	B4(nb) IS 21.7 × 1.814 RH fem	C(nf) IS 21.7 × 1.814 RH fem	UNI 4412 W24.51 × 1.814 RH fem

Abbreviations: nb Belgian standard; DIN German Standard
nf French standard (AFNOR); RU, RI Dutch Standard
ns Swedish standard (SIS); UNI Italian standard
ne Spanish standard BI; British standard (BSI)

Table 2.2 European Valve Connections (b, Continued).

Gas	DE/AT	CH	DK/SE	NL	UK
Air	DIN 9 /DIN 6 /DIN 13	SN219505/10	DIN 13/DIN 6	RU-1	BS341 No 3
	W26.44 × 1.814 RH fem	G5/8" male	W22.92 × 1.814 RH male	W21.8 × 1.814 RH fem	G5/8"-14 RH
	W21.8 × 1.814 RH fem	PI 219507–2.9	W21.8 × 1.814 RH fem		
	G5/8"		=W21.8 × 1/14m		PI: Pin Index
Argon	DIN 6/DIN 10	SN219505/7	B(ns) = DIN 10	RU-3	BS341 No 3
	W21.8 × 1.814 RH fem	W21.8 × 1.814 RH fem	W24.32 × 1.814 RH fem	W24.32 × 1.814 RH fem	G5/8"-14 RH
	W24.32 × 1.814 RH fem				PI: Pin Index
Carbon Dioxide	DIN 6	SN219505/7	A(ns) = DIN 6	RU-1	BS 341–8/PI
	W21.8 × 1.814 RH fem	W21.8 × 1.814 RH fem	W21.8 × 1.814RH fem	W21.8 × 1.814 RH fem	0.860" × 14TPI (M)
		PI 219507–2.3			PinIndex ISO407
Helium	DIN 6	SN219505/7	B(ns) = DIN 10	RU-3	BS341 No 3
	W21.8 × 1.814 RH fem	W21.8 × 1.814 RH fem	W24.32 × 1.814 RH fem	W24.32 × 1.814 RH fem	G5/8"-14 RH
Nitrogen	DIN 10	SN219505/8	B(ns) = DIN 10	RU-3	BS341 No 3
	W24.32 × 1.814 RH fem	W24.32 × 1.814 RH fem	W24.32 × 1.814 RH fem	W24.32 × 1.814 RH fem	G5/8"-14 RH
Nitrous Oxide	DIN 11/DIN 12	SN219505/9	G(nf)	RU-1	BS 341–13/PI
	W16.6 × 1.336 RH fem	G3/8"	IS 26 × 1.5 RH male	W21.8 × 1.814 RH fem	11/16" × 20 TPI (M)
	G3/4" male (<3L)	PI 219507–2.7			
					PinIndex ISO407
Oxygen	DIN 9	SN219595/2	A(ns) = DIN 6	RI-2	BS341-3/EN850 (PI)
	W26.44 × 1.814 RH fem	G3/4"	W21.8 × 1.814RH fem	W22.92 × 1.814 RH male	G5/8"-14 RH/PI
	G3/4"	PI 219507–2.4			
Xenon	DIN 6	DIN 6	B(ns) = DIN 10	RU-3	BS341 No 3
	W21.8 × 1.814 RH fem	W21.8 × 1.814 RH fem	W24.32 × 1.814 RH fem	W24.32 × 1.814 RH fem	G5/8"-14 RH

Abbreviations:
nb Belgian standard
nf French standard (AFNOR)
ns Swedish standard (SIS)
ne Spanish standard
DIN German Standard
RU, RI Dutch Standard
UNI Italian standard
BI British standard (BSI)

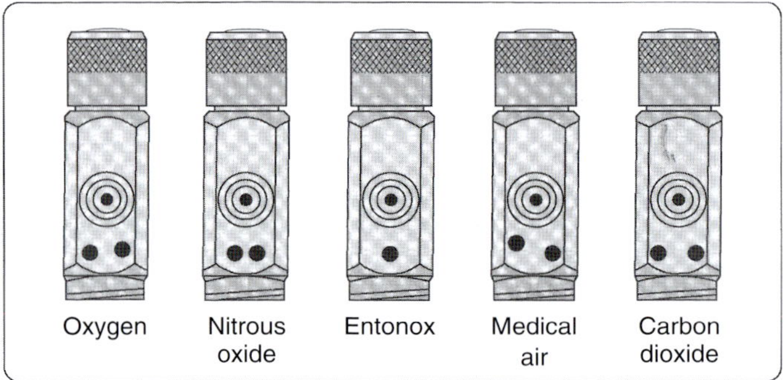

Figure 2.9 Pin-index-type valves [49].

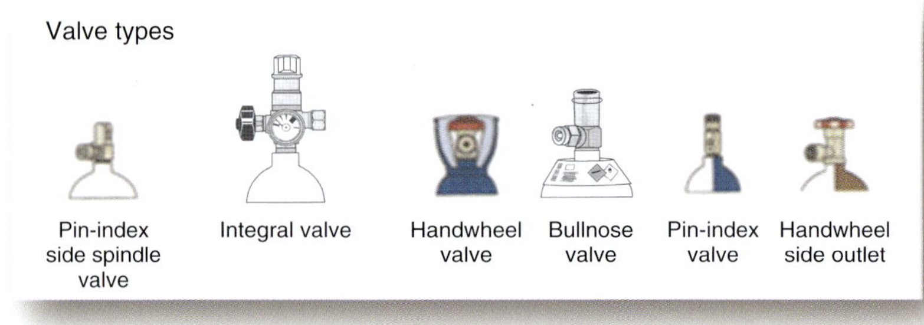

Figure 2.10 Silhouettes of different valve types [49].

cylinder side. Figure 2.9 shows the different positioning of the holes to fit with the appropriate connection of the regulator.

For a quick identification of the different types of valves, refer to the figure 2.10.

2.1.6.1 Accessories for Valves: Gaskets

Coming along from the original technical valve, two main points have to be considered: the valve is fixed in the cylinder with a conical thread; in Europe two diameters 25E and 17E are commonly used for cylinders ≥5 l and cylinders ≤5 l, respectively.

Materials for the gasket have to be resistant to the oxygen at high pressure to avoid ignition of the valve or the valve thread. To make this connection tight for a longer period of time, in the early days, either mineral suspensions (after baking out to establish a tight connection) or a tin beaker in the shape of the valve thread were used and screwed with the valve together into the cylinder. At that time, it was necessary to check the neck blower after filling, to sort out cylinders with unreliable gaskets.

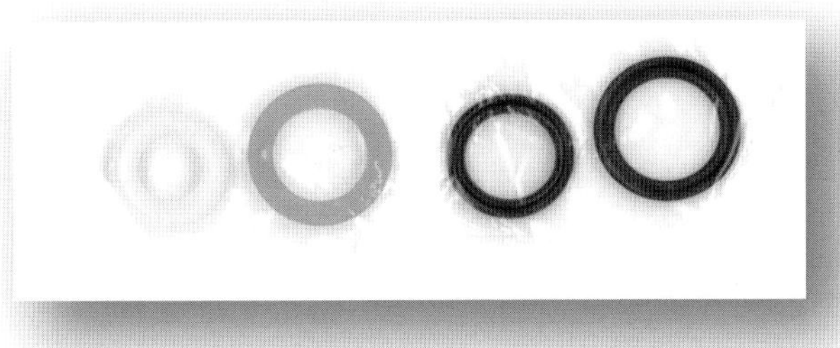

Figure 2.11 Typical washers for different gases, made from different materials, packed in plastic to avoid contamination with grease.

In fact, the cylinder connections in front of the regulator are crucial for the tightness of the cylinder: if there is a leak, the gas inside the cylinder would be vented rapidly, owing to the 200 bar pressure. Under the conditions in a hospital this would mean that the gas is vented into the atmosphere before it reaches the patient and the cylinders would get empty very quickly. For common systems (without pin-index) either a washer or an O-ring is to be used, depending on the layout of the connection of the regulator. If this is to be fixed with a wrench, a washer must be used, if it is to be fixed by hand, an O-ring has to be used. Depending on the layout of the construction of the connector only one of the two is possible: O-rings are pressed into a chamber to reach full tightness, and can never be changed into a washer! This would result in a leak.

Even if gaskets do not look very different (see Figure 2.11), gaskets for oxygen have to be kept separate and kept oil- and grease-free to avoid any danger of flash fire with oxygen. In fact the cylinder valve and the application device (e.g. the pressure regulator) can be regarded as a unit, every part of the unit has to be thoroughly constructed, built and tested to safely prevent flash-fire under oxygen service. In Germany as an example, all parts in oxygen service have to tested for flash-fire by BAM (Bundesanstalt für Materialprüfung), a federal institution.

The other point, even with simpler built industrial valves is the tightness of the valve itself. Here, usually the spindle seals the gas flow with an O-ring in its simplest form or a variety of geometric inserts with appropriate materials to provide tightness. Again the condition is that the construction must be oxygen-proof, that is, the possibility of flash fire or ignition in the valve has been tested according to the valid standards. Figure 2.13 shows a very simple straight-on construction, originally used both for technical and medicinal gases.

2.1.6.2 Valves with Integrated Residual Pressure/Nonreturn Cartridge (NRV/PRV)

Taking into account the experience of the last 50 years, in most cases, the inner surface of the cylinder remains unaffected, as it is only in contact with the pure

Figure 2.12 Valve with RPV/NRV cartridge to avoid contamination during filling (own picture).

media in ordinary service. A standard error situation is the complete emptying of the cylinder and leaving the valve open for the time period of returning to the filler. This led to an extended accessory: the valve was equipped with a so-called RPV/NRV cartridge at the outlet (RPV, rest pressure valve; and NRV, nonreturn valve) (Figure 2.12).

This type of valving of a cylinder safely prevents the common mistake that could contaminate the internal cylinder surface (namely, small cylinders): either by complete venting of the cylinder or by leaving the valve open. Here it might result in inflation of ambient air into the cylinder owing to temperature changes, creating a "breathing" cylinder. As ambient air contains considerable amounts of water, this would lead to a superficial rust of the interior surface of the cylinder.

2.1.6.3 **Integrated Valves**

For the ordinary valves (Figure 2.13) used in connections, a big wrench was needed to make the connection tight. By changing the flat gaskets in the ordinary connections to O-rings in the medicinal connections, they became attachable by hand, without a wrench.

It is clear enough that the hospital environment is not a good place to perform critical connections such as those of oxygen cylinders in this manner. It takes time and care to attach and flash fires can destroy both the cylinder neck and pressure regulators. To avoid the mounting of a pressure regulator,

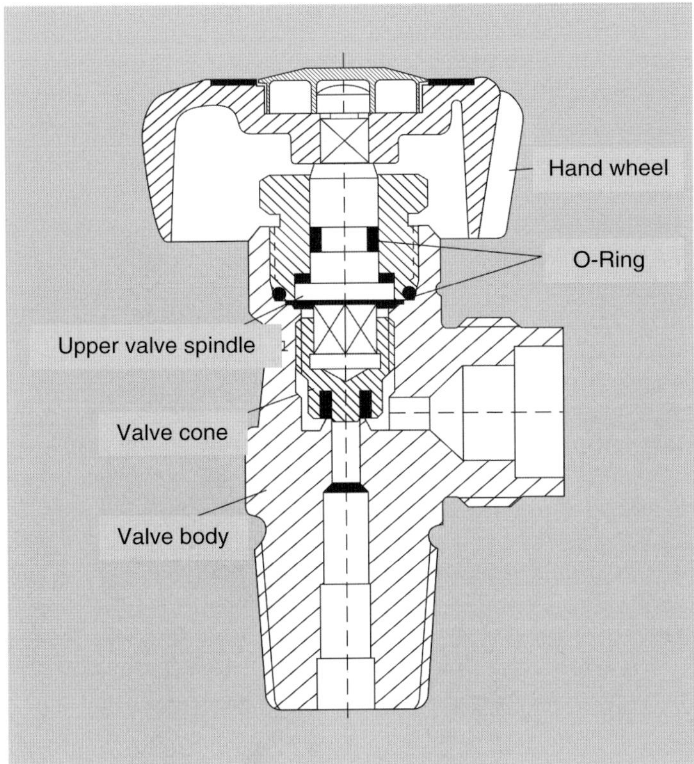

Figure 2.13 Ordinary standard valve [51].

valves with an integrated pressure-regulating device were developed and, in the last couple of years, have been replacing in some countries the simple cylinder valves. The advantages of this construction are the ease of handling and the availability of the gas, without the mounting of any external device. Figure 2.14 shows a typical integrated valve with an integrated pressure regulator unit.

Integral valves combine ease of use for the customer in the hospital and separate filling connectors for the gas company. The customers are able to get the gas at the appropriate low pressure without further pressure regulation. Only a moistener, to moisten the gas, has to be included in the setup, before the gas is administered to the patient. A further simplification and additional safety feature is the ready-mounted valve guard, which cannot be misplaced or forgotten.

In state-of-the-art designs, the integrated valve has been equipped with digital meters and indicators to show the amount of gas remaining in the cylinder (Figure 2.15).

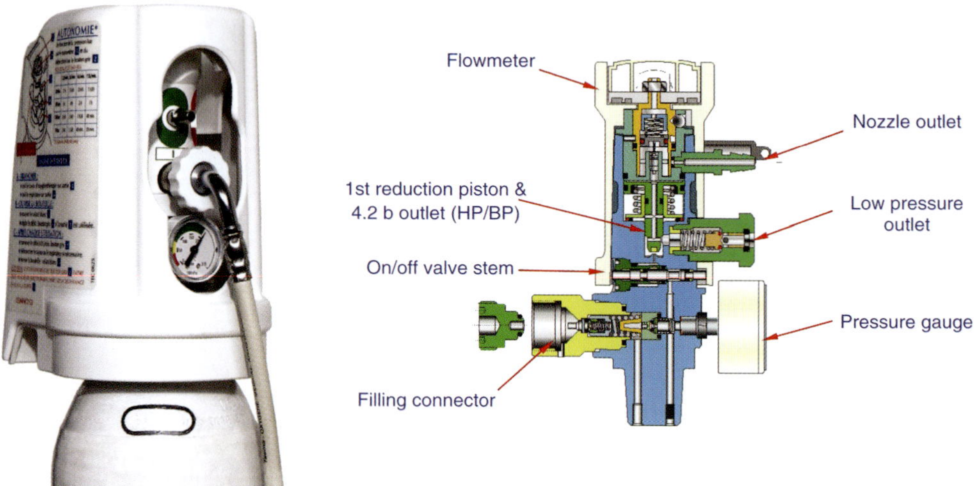

Figure 2.14 Integral valve (Air Liquide, [51]).

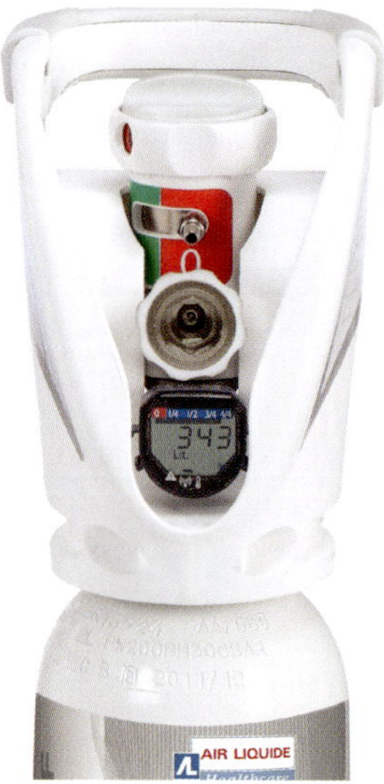

Figure 2.15 Integral valve (TAKEO®) with digital indicator for the residual content in the cylinder (Air Liquide, [52]).

2.2 Non-transportable Pressure Receptacles: Stationary (Pressure) Tanks for Cryogenic Liquids

These containers are equipped with suitable wall isolation to reduce the transport of energy from the outside of the container to the inside. Usually, a two-walled construction is used; the intermediate space between the two walls is evacuated and filled either with isolating materials, such as Perlite® or a specified wrapping.

For an almost quantitative suppression of energy transport, the container is wrapped with sealing foil aluminated on one side. A good measurement of the performance of such a container is the vaporizing rate, the time ratio of half or total vaporization of the stored media.

2.2.1 Safety Measures on Stationary Tanks

To avoid unwanted blowup, all liquid-carrying pipes or parts are secured with safety valves. In case of improper pressure rise, these safety valves would release the pressure before the container can burst. In some countries of the EU special devices prevent overfilling of the tank during the refill-operation: dead-man switches limit the running of the pump to the time when the driver stands aside, pressure switches switch of the pump in emergency case, e.g. when the resistance exceeds predefined limits (for example when the tank is short before overfill) etc.

Attached to the tank is a local vaporizer that helps maintain a constant pressure in the tank. This pressure is adjusted in such a way that it is enough for the customer, depending on the consumption of the user/hospital.

For economic reasons, transportable cryo-tanks (e.g. tank trucks) sometimes have to change their service gas. The procedure for requalifying the container for medical gases should be well documented as a written procedure that is followed with appropriate analyses before release.

Liquefied cryogenic gas is stored under atmospheric pressure, to allow the gas to vaporize without any danger for the environment owing to the vast volume expansion by vaporization of the liquid. To withdraw liquid from the container, a siphon-type principle is used (see Figure 2.16): by opening valve *x*,

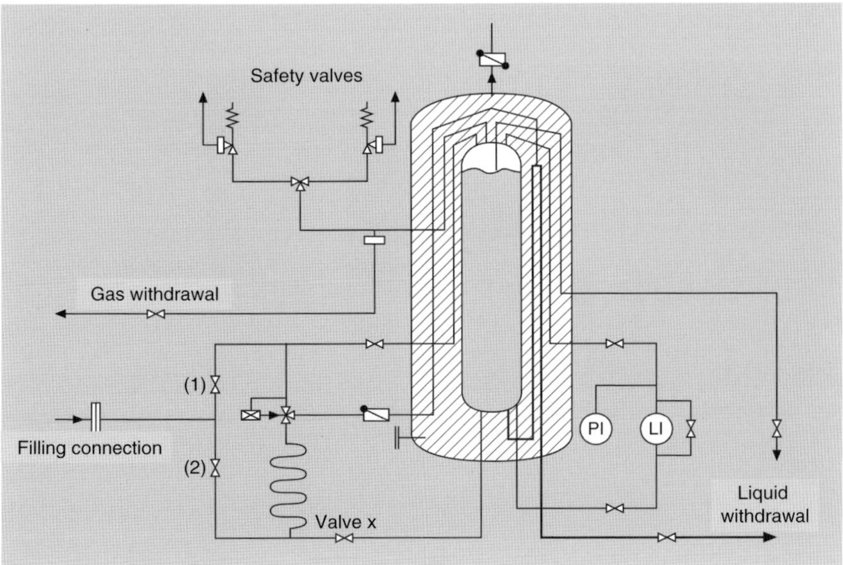

Figure 2.16 Stationary container for cryoliquids [14].

a constant pressure builds up. The liquid leaves through the outlet the container and can be used.

2.2.2
European Pharmacopoeia View on Cryo-Tanks as Containment for Drugs

From the viewpoint of the pharmacist, the cryo-container is special because of two major factors: the used materials can vary from almost simple (but cold-resistant) steels to stainless steels for inner (product-holding) containers. Following good practices in the food industry, it might be recommended to use stainless steel as material for the product container. In principle, other steel would be sufficient, but then internal rust cannot be excluded during operation time of the tank. The rust, although not directly dangerous to health, would result in particles coming out of the container along with the oxygen flow. That is why, the inner container should be made from stainless steel to avoid any particulate load of the gas being withdrawn. The container should never be emptied totally for energetic reasons; if the tank is refilled, the liquid added during refill will dilute the remaining content in the tank.This procedure seems to be generally accepted.The supplier has to take care to use only released product for the refill, to ensure that the quality in the stock does not decrease; this is usually achieved by the supplier, who takes suitable precautions (Figure 2.17).

As for transportable pressure vessels (Figure 2.18), stationary cryo-tanks also have not been found so far to have a mentioning in the Pharmacopoeia.

Figure 2.17 Truck refilling a tank [50].

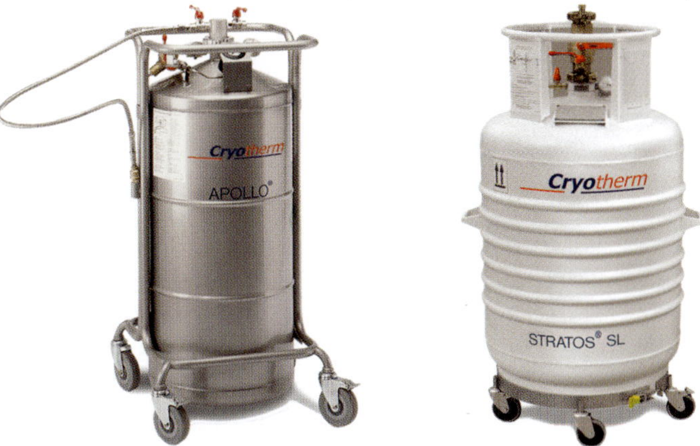

Figure 2.18 Transportable mini-container for cryoliquids [14].

2.2.3 Inner and Outer Surfaces of Cryo-Tanks

The other topic important for the pharmacist is the way to operate the tank: owing to contractual reasons in Europe, a common stationary tank is seldom operated longer than 5 to 10 years at a certain customer. After that time, the tank is changed for different reasons and replaced by another tank of similar size. The new tank must be refurbished and must come from other customers after refurbishment.

It is obvious that refurbishment and other treatments should be documented thoroughly and the documents have to be available for the release of the tank for medicinal purposes, by the qualified person.

The methods of maintaining storage tanks without contamination by air or other gases need to be fixed as also the tests before a tank is released to enter into the supply chain again.

As the big cryo-tanks reveal no possibility for a classical internal inspection, for example, by opening a manhole or a similar hatch, all efforts must be taken to prevent contaminations from entering the tank. The materials of the tank (especially the inner container, which is in contact with the medicinal product) must have been chosen in a way that is suitable for this special type of use of the tank. Water is the most hazardous contamination in this case, so it might be recommended to store tanks before refurbishment under slight inert gas pressure.

A good check of this objective is the measurement of the purity of the purging gas when the tank is at the refurbisher. If oxygen or water is present in the purge gas, this might be an indicator of contamination.

If water and air can be excluded from entering the tank, the inner surface can be kept clean and free from any corrosion. A gas tank should be used just for one dedicated gas. If necessary, it is possible to change the gas service under controlled conditions; again, the procedures on how to purge with the new gas, how to make an analysis, and how to make the release should be well documented and strictly followed.

2.2.4
Accessories for Cryo-Containers

Accessories for cryo-containers have to be laid out to withstand contact with the cryogenic media. All connecting piping and valves can be wetted by the cryogenic liquid and are thus subject to cold embrittlement if they are not made from copper or cold-resistant steels.

For the operation of the containers one should order original spare parts from the supplier and leave all service and maintenance operations to experts. There are a number of safety devices necessary on liquid tanks, which are similarly used all over the industry.

The accessories as such are (safety) valves, all operable under cryo conditions, and functional parts, such as the vaporizer, the pressure manifold, the valves, and connecting pipes (Figure 2.19). The exhaust pipes of the safety valves should not be directed to the tank itself or toward where people could pass. Vented gas has almost cryogenic temperature and can damage the tank or can be hazardous to the people in the vicinity.

The withdrawal of the cryogenic liquid is managed via an extra piping if necessary; usually, the tanks are equipped in a way that all liquid passes the appropriate vaporizer before leaving as gas flow and entering into the piping of the hospital. Withdrawal of oxygen is, from the legislator's viewpoint, a production process and thus complicated with further procedures along the GMP.

Typically, hospitals have continuous consumption of oxygen, with the tank delivering constant amounts of vaporized gas over a long period of time, often

Figure 2.19 Safety valves at a cryogenic tank [50].

Figure 2.20 Ice coating on a vaporizer.

interrupted only by necessary revisions of the tank or its exchange. Over the year, the vaporizer faces different meteorological conditions, so in some cases, in middle Europe usually in winter, the piping between the tank and the vaporizer itself is collects a coating of ice, sometimes of considerable thickness (Figures 2.20 and 2.21).

Although the ice coating does not affect the function of the tank (provided no operational valves have become inoperable because of the coating), it should be removed from time to time, as the slowly growing ice could damage the piping by bending it out of the shape. The method of choice is not to hammer away the ice (as all parts are under pressure), but to take an industrial blow-dryer to thaw away the ice coating.

When the tank is loaded from a tank truck, a common mishap that can occur when either the safety valves of the tank are blowing or the breakdown of the supply pressure to the hospital occurs, which gives rise to an alarm in the central medical gas supply and sometimes leads even to switching to emergency feed-in. Both occurrences have something to do with the design of the inlet pipe into the hospital: if this is of a comparatively small diameter, and the extension of the pipeline system has been considerably enlarged during the years, often the inlet pressure has to be raised, to get enough oxygen into the network.

Figure 2.21 Ice coating on a valve box.

In fact, this means that the operation pressure of the tank will be raised through the years, finally ending up with a pressure close to the limiting pressure of the safety valves. The lower limit of the pressure sensor to detect oxygen deficiency and to start the emergency feed is only a few bars below the higher limit of the pressure and will give an alarm immediately. When the tank should be refilled, the driver has to operate the pressure between these two limits, to prevent safety valves from opening or to avoid the switch to emergency feed.

2.2.5 Choice of the Good Location for Tanks

The choice of a good location for a liquid oxygen tank of a hospital often is very difficult, as any arbitrary place cannot be chosen.

First of all, the tank is used as storage for a couple of tons of liquid oxygen. During ordinary operation, the tank emits cold oxygen through the safety valves, collecting ice at cold tubes near the valves, and emits sounds. The tank is frequently refilled by a medium to large sized tank truck with the emission of noise and possibly blocking the access to traffic; additionally, the valves have to be moved during refill and oxygen is vented.

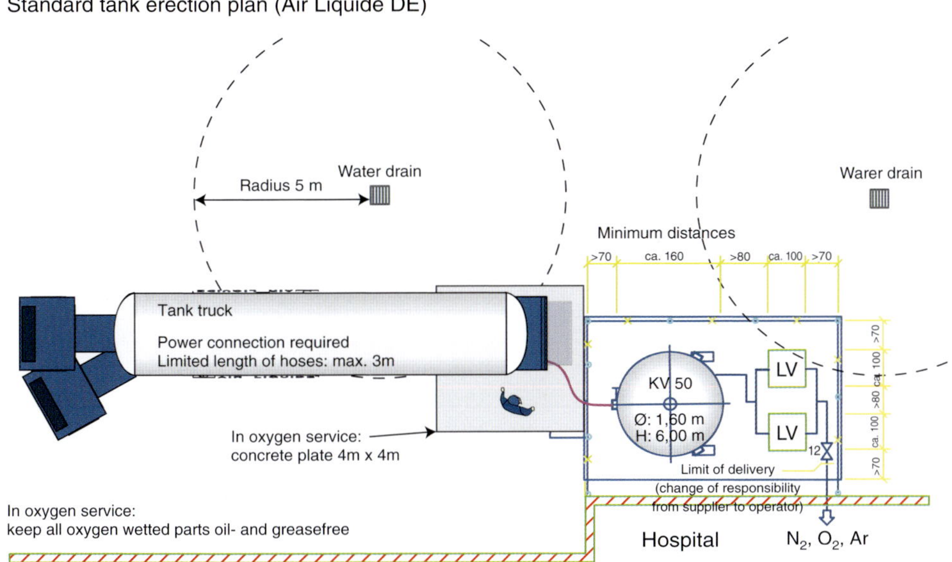

Figure 2.22 Minimum distances for safe positioning of a cryogenic tank [50].

With respect to the maximum size of the tank truck (44 t), only dedicated streets can be used in the area of a hospital. Additional space is needed for the truck maneuvering.

All this limits the available location, even more, if the architect did have special ideas when positioning the tank in the hospital's courtyard. The tank itself requires some minimal distances to gullies and to flammable materials, and needs special materials for the ground on which it rests. Figure 2.22 shows the most relevant features for the position of a cryogenic tank (Lit).

2.3 Medicinal Gas Pipeline Systems (MGPS)

As we have seen on the previous pages, although light and breezy, the gases exhibit special problems of transport. To supply single patients with gas, it is either necessary that each patient breathes from his own cylinder, or we should have installed a branched network in the hospital, supplying oxygen from a socket in the wall of the patient's room.

Sure enough, such an installation resolves a whole lot of other difficulties, but we are today at a point where we can use all the experience collected in the past, which has been precipitated in the form of a number of standards describing the design and the operation of a MGPS.

The MGPS is also the missing link between the tank and the vaporizer on the one hand and the hospital inlet on the other. This inlet branches into several buildings, levels, and a variety of rooms, from the operation theater to the simple patient's

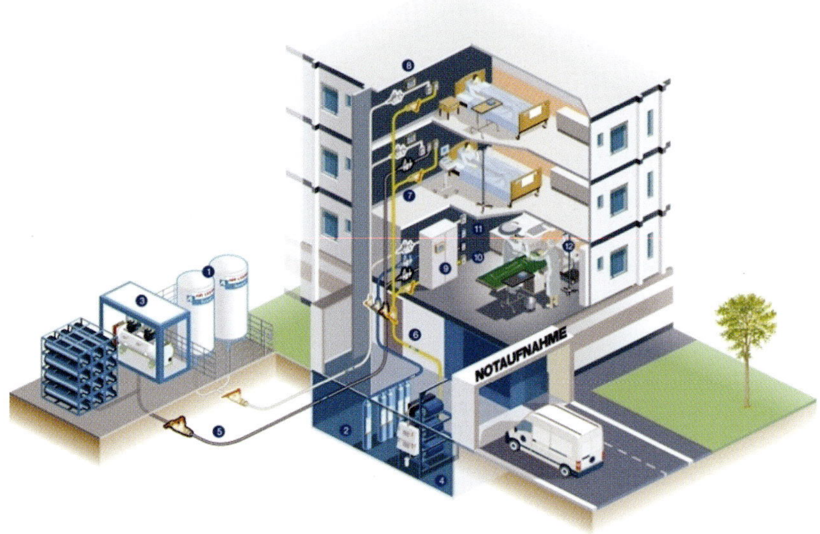

1	Stationary oxygen tank (double)
2	Feeding of MGPS (in the cellar)
3	Compressors to produce medical compressed air
4	Vacuum generation pump, vacuum piping system (EN ISO 7396-2)
5	Copper piping (specified for medicinal use)
6	Shut-off valves
7	Second stage pressure regulation
8	Alarm system: indicator
9	Safety cabinet
10	Wall sockets for medicinal gases
11	Wall socket for medical air
12	Collection of excess anesthetic gases

Figure 2.23 Scheme of an MGPS in a hospital (Graphics: Air Liquide).

room (Figure 2.23). In this book, we only highlight some aspects of the standard and the do's and don'ts of operation of an MGPS. The operation is a very interesting task, as we find at least four parties carrying jointly he responsibility for the proper function of the MGPS: the pharmacist, the responsible house technician (operator of the MGPS), the nursing staff, and finally, the patient.

For a better view of the requirements, we should start with the patient: if constantly connected to the gas supply, he becomes addicted to that form of receiving oxygen. If the piping transports contaminated gas or even worse, other gas than oxygen, the patient is doomed to death. So the basic principle is for the responsible operator to avoid any gas other than the one nominated to enter the pipe and keep contaminations within the limits of the pharmacopoeia (to keep on measuring!)

In the presence of pure oxygen, all materials can burn down. The ignition can be induced by metal powder or by other small particles still present in the pipe from numerous sources: abrasion from valves, remainder from the last mounting,

(a)

(b)

Figure 2.24 Influence of forming gas on the inner surface of the pipes (a) good quality and (b) poor quality).

detachments by hydraulic shocks, and so on. If the piping contains materials that form toxic products under reaction with oxygen, these materials will be transported within a split second to the patient.

In the literature, we can find incidents where hydrogen fluoride has been formed by the ignition of fluorine-containing gaskets, intoxicating numerous patients at the end of the pipe.[4)]

Typically, medical gas pipes consist of copper tubes of special quality. The copper tubes are connected by hard soldering, either with hydrocarbons (acetylene) or hydrogen as fuel gas. The heating of the pipe to dark-red glow requires the securing of the inner surface from oxygen by an oxygen-free forming gas (e.g., nitrogen) to avoid the generation of copper oxide, a dark black powder. If the connections of the pipes have been soldered without the regular use of forming gas, the connections will release particles for the remaining part of the life of the piping. Figure 2.24 shows two examples, one for a good connection, soldered under forming gas, and the other, soldered without forming gas.

It is obvious that the contamination with CuO-powder on the bad soldering is source for continuous particle contamination, which must be avoided to meet the specs of EN ISO 7396-1. The provisions to be made during design and assembly of a gas supply system for medicinal gases are described in detail in the EN ISO 7396-1 [53]. The same standard gives the framework for testing after the assembly and for the operation. As the requirements and the approach are extensively characterized there, we confine ourselves to a view of the principles behind the device, to make the observer or the latter user aware of the inherent difficulties or risks that accompany unprofessional use of such a device.

Truly, the MGPS of a hospital is a medical device, thus underlying the regulations according to 93/42/EC, last amendment 2007/47/EC. The operating unit thus has the full responsibility that the medical device is in the right condition to transport gaseous drug from the feeding of the MGPS to the outlet at the patient's

4) Using of halogen-containing washers and gaskets has been banned by several industrial gas companies and the EIGA. In documented cases, these materials had ignited under oxygen pressure, the combustion products being HF or HCl, depending upon the decomposed material. Highly toxic HF and HCl are transported with the gas flow to the wall outlets, finally intoxicating the patient.

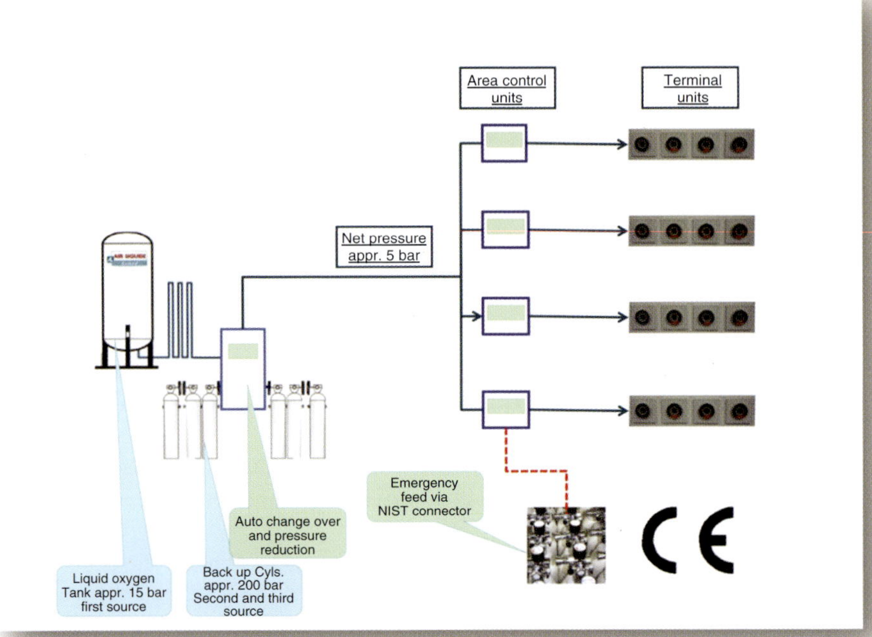

Figure 2.25 Key Elements a Medical Gas Pipeline System (own graphics).

wall socket. The operator is responsible for routine checks and quality management of the medical device, the pharmacist remains responsible for the pharmaceutical quality of the medicinal gas.

2.3.1 Elements of a Medical Gas Pipeline System (MGPS)

The general design of a medical gas supply system (MGPS) is not very complicated: starting with the feed-in (e.g., gas from a cryogenic oxygen tank, from cylinders or from an air compressor) the gas is led through the piping at medium pressure to the different buildings, where it is further branched to the different levels, and from there to the different rooms. Often, national principles of operation are applied here.

- *Source*: This could be an oxygen tank or a bundle or an assortment of cylinders, all connected to the feed-in of the MGPS. According to EN ISO 7396-1, the source must be tripled, to avoid a stop in the gas flow. Depending upon the country, different solutions are offered by the gas industry: Tank – Bundle – Cylinders/Tank – Tank – Cylinders/Tank – Cylinders – Cylinders/ Cylinders – Cylinders – Cylinders or similar combinations (Figures 2.25 and 2.26).

Figure 2.26 Typical sourcing, picture shows the backup feed with cylinders.[5)]

- *Pressure regulation:* A set of pressure regulators are used to reduce the pressure (usually, in two stages from the cylinder: 200 bar feeding from cylinders, first stage approximately 15 bar, second stage 5 bar). If the gas comes from the tank, it is already controlled for the first stage with 15 bar (Figure 2.27).
- *Piping:* Piping is made from copper tubing and valving is arranged in such a way that the building, the levels, the wards can be separated without interfering with other buildings, levels, or wards (Figure 2.28).
- *Area control units or control closing boxes:* Area control units contain the valves to shut off areas from the supply. This is necessary when leaks have to be detected or when installations have to be carried out. Usually, they also contain a connector (NIST or DISS) to connect an external source to the piping area (Figure 2.29).

Area control units also contain a service connection and shutoff valve. In certain cases, the connection to the central supply can be shut off with the valve, and gas can be fed through the service connection into the piping via a NIST or DISS connection.

Figure 2.29 shows such a unit.

2.3.1.1 Gas Terminal Units (Wall Sockets)

The gas terminal provides a gas-specific connection and the shutoff of the pipe (Figure 2.30). If a hose is plugged in via a matching connector, the gas flow is

5) Design and pressure regulator unit by Draeger.

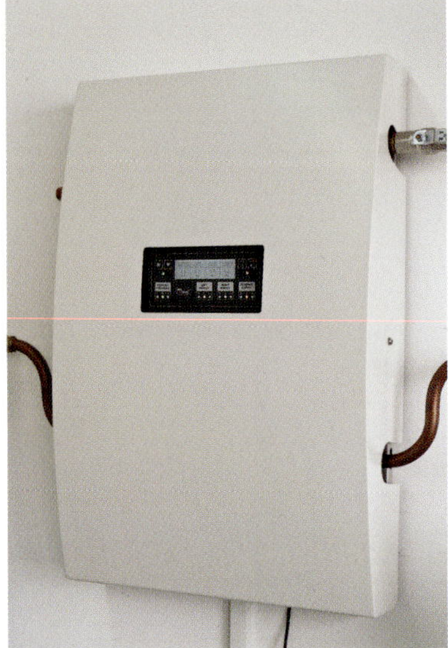

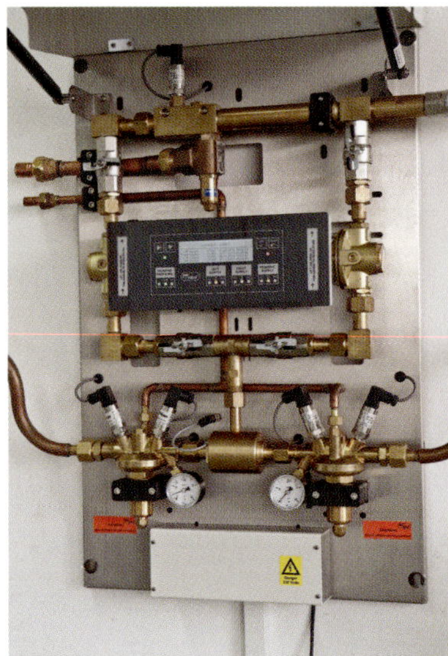

Figure 2.27 Feeding unit from the outside/with open cover.[6)]

Figure 2.28 Copper piping in a hospital (during construction).

opened and the gas starts flowing. However, the wall sockets in all European countries are different for different gases; there is no European standardization on this aspect. The wall sockets automatically close when the appliance is demounted. The red label on the right unit is a warning sign, not to use the gas before release of this part of the piping.

6) Design and pressure regulator unit by Medicop, Medicop d.o.o., Obrtna 43, SI-9000 Murska Sobota, Slovenia.

Figure 2.29 Shutoff valve and NIST connection in area control unit/control closing box (6).

Figure 2.30 Terminal units (German design, six corners for oxygen, four corners for medical air).

2.3.2 Tests and Checks before Going Onstream

When erecting an MGPS in a hospital, EN ISO 7396-1 prescribes all necessary actions to design and to put on stream the device onstream according to experience and good practice. As the whole system is a medical device, the conformity has to be stated according to 93/42/EC by a qualified person (which is not a pharmacist but rather the technical head of the supplier of the MGPS). Before doing so, the person has to perform and document a number of tests as specified in the standard:

All tests have to be performed and documented by the producer of the MGPS

a) before **covering** of pipings
   - check of markings on the pipe and pipeline supports
   - check of conformance with design specifications
b) Tests, checks, and procedures before **going on stream**
   - test for leaks and mechanical integrity
   - test of area shut-off valves, leakage, and closure
   - test for cross-connections
   - test for obstruction and flow
   - checks of terminal units and NIST or DISS connectors for mechanical function, gas specificity, and identification
   - tests or checks of system performance
   - test of pressure relief valves
   - test of all sources of supply
   - tests of monitoring and alarm systems
   - test for particulate contamination of pipeline distribution systems
   - tests of the quality of medical air produced by air compressor systems purging of the system with the appropriate (specified) gas
   - check of the identity of the gas
c) Documentation (acc. EN 1041 or equivalent national standards)
   - Instruction for use
   - Operational Management Information
   - "As-installed" drawings
   - Electrical diagrams.

All headlines are explained in the standard and specimen of used forms and documentation sheets are also given.

2.3.3 Operation of a Central Medical Supply System

The operation of a MGPS system is a complex task. The responsibilities should be distributed among different persons, each of them responsible for his part of the actions. ISO 7396-1 defines eight persons with a key responsibility:

Persons with key responsibility for MGPS operation

EM	Executive Manager
FEM	Facility Engineering Manager
AP	Authorized Person
CP	Competent Person
QC	Quality Controller
DMO	Designated Medical Officer
DNO	Designated Nursing Officer
DP	Designated Person

By intelligent announcement, the different roles can complement one another so that the operation of the MGPS is facilitated.

One of the major requirements for the operation of an MGPS is the risk analysis procedure. In case of an error, the whole or parts of the MGPS can fail. It is of utmost importance to have a thorough planning for the different severities of such a case.

We might look into the EN 7396-1, the recommended emergency time period where the MGPS should be working and supplying oxygen or medical air to the patients, even if the primary supply has failed, is set to a minimum of 24 h. In this time, the emergency communication chain to the gas supplier and his emergency system should be indented and working, so that either liquid supply can be restarted or the supply is secured with cylinders.

Depending on the failure, different strategies must be developed for the different cases: in case of no liquid supply (e.g., main roads closed owing to weather conditions, accidents, or extraordinary traffic jam), parts of the MGPS should be separated and local feeds can be used except for intensive care units or operation theaters. In other cases, the supplier should have a maintenance service in due time at the place either to repair a defective part or to put an emergency tank at the place.

In the case of shortcomings of cylinders (e.g., due to major catastrophes, earthquakes, or airplane accidents) the supplier has to be contracted to have in due time specific numbers of filled cylinders and of specified size at hand to support the hospital. For this case, the supplier has to undertake long-term planning, as in the routine case the distance between the filling plant and the hospital might reach some hundred kilometers.

In every case, the hospital would be well advised to discuss emergency questions and possible emergency scenarios right before they occur, to secure the reliable supply of oxygen and medical air.

2.3.4 Maintenance and Service, Pharmaceutical View

Once the MGPS has been passed all tests and the certification by the supplier to be EU-conforming has been done, the MGPS needs routine checks and maintenance: the MGPS as a whole and all included elements should now bear the CE-mark.

For service, one has to decide whether the CE-mark is violated. This would be the case, for example, when opening pipes or exchanging parts. All these activities have to be carefully registered and documented to be traceable with regard to changes or other interventions into the closed system.

In all cases of doubt, the authorized person has to send an alarm to the supplier or the notified body to refresh the CE-mark.

Looking at the requirements of the patients, it is obvious that to avoid any harm to them, a sophisticated lock-out/tag-out procedure should be in place. Parts of the MGPS should only be separated from the feed under supervision of the authorized or competent person, ensuring that all wards affected have been informed before about a breakdown of the gas supply and secured their supply by other measures.

Going back on stream in the same way needs a release of the authorized person, usually taking place after the necessary quality checks have been positively performed. Here, it is of particular importance that they make the checks at the end of the piping, at the terminal units. Impurities in the pipes do not mix with pure gas, they are pushed "*in toto*" through the piping, which makes them dangerous. In the small diameters of the piping in the wards, such a gas bubble can take a few minutes to pass, that is a time period, in which a patient would not survive, if the gas bubble consists of more than 90 vol% of nitrogen. In every case of a failure of the MGPS, fatalities become highly probable.

As already mentioned, from the technical requirements, the pipes have to be purged with nitrogen during soldering to avoid particle emission from badly soldered seams. On the other hand, it must be absolutely clear to have all nitrogen replaced by oxygen, before the pipeline system or even parts of the pipeline system go onstream to the patient. This can only be proved by using the appropriate techniques and strategies for purging and for checks by oxygen measurement.

Maintenance works are nowadays often subject of outsourcing, as the workers will be needed only from time to time. Using external companies for maintenance works should be assessed with Quality Risk Management methods to be sure to have all necessary risk mitigation measures met: Work-permit procedure, lock-out/tag-out, thorough check of all workers (qualifications, work planning and documentations) ready for start-up review, final release for going onstream by the responsible person. Often the vacuum piping is regarded as part of the MGPS, as the same people often service it. It is obvious that a vacuum pipe does not transport purity: in spite of all precautions met, germs and debris often enter the vacuum lines, raising special danger of contamination to the service people, when changing filters or valves of the vacuum line. For the responsible operator mainly the risk of diversion is relevant: unprofessional handling of the contaminated filters leads to infection of the service people or the service people contaminate the MGPS with germs from the vacuum line.

3
Analytical Methods for Gases (as Described in Ph. Eur.)

3.1
Sampling

Correct sampling of gases is most important in order to avoid invalid analytical results by discrimination of trace impurities. In view of collecting a representative sample of a gas, the different physical states have to be distinguished: permanent gases, gases liquefied under pressure, and cryogenic gases.

3.1.1
Permanent Gases

These gases are stored in pressure receptacles (steel or aluminum cylinders). The high pressure in the cylinders requires the use of a pressure regulator to avoid damage to the analytical instrumentation and for the safety of the sampling analyst. The content of gas in the cylinder is in a direct ratio to the pressure of the cylinder, for 'ideal' gases (gases that behave along the thermodynamic laws) this is a linear function (Figure 3.1).

If the cylinder or receptacle contains a mixture, the distribution of the gaseous constituents has to be checked to be homogeneous.

It is part of the thorough training for the lab-staff to be able to distinguish between actual permanent gases and gases with changing state of matter, near the condensation or critical points. If it turns out that at least one of the components might have condensated, it has to be checked if this is due to a considerable temperature choc during transport or if the gases in the cylinder are at ambient temperature in the vicinity of their boiling points. Samples can be taken either into another appropriate (i.e., pressure-proof) sample unit or direct into the instrumentation (see Figure 3.2) via stainless steel pipes or thick-walled Teflon® tubes.

If the sample is to enter directly into the instrument, it would be appropriate to have sufficient bypass venting to avoid any choking in the sample line. Otherwise, one would receive unpredictable and unrepeatable pressure effects disturbing the flow and thus the signal of the compound to be estimated, as most of the instruments used are sensitive to the mass flow passing the probe.

Medical Gases: Production, Applications and Safety, First Edition. Hartwig Müller.

Permanent gas under pressure

- Reading of the gauge add the residual quantity in the cylinder are in a fixed ratio (proportional). Example: 10L cylinder, exact values only for ideal gases

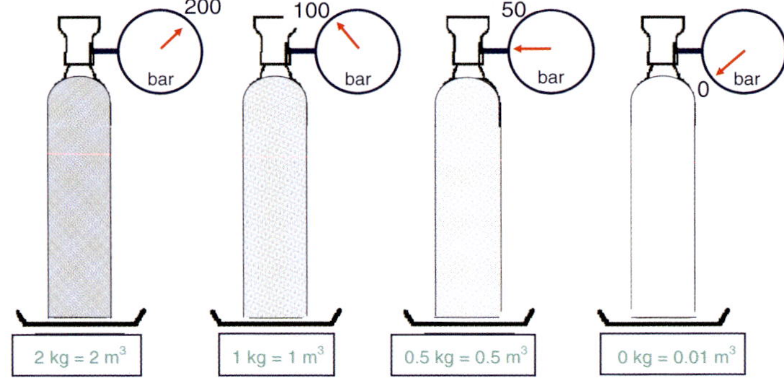

Figure 3.1 Gas cylinder with a permanent gas.

If plastic or Teflon® tubes are used to connect the sampler and the device, they must clearly resist the applied pressure, as well as the connection system used. Depending on the nature of the plastic or Teflon® tube, diffusion of water from the air into the tube walls and finally into the gas can occur.

When a container is used to collect the sample (e.g., a small steel or aluminum cylinder), it has to be ensured that it can receive enough sample to carry out all necessary analyses later on. To choose the right size of the sample container, the amount of gas inside has to be calculated: the volume of the container has to be multiplied by the pressure of the sample line minus the volume of the cylinder, as the container can only be emptied until pressure equilibrium.

The size of the sample cylinder should be appropriate to yield enough of the sample, particularly when the pressure is low. One must be wary if the pressure is too low (this would require a sampling container of large volume and result in low flows due to minor pressure differences). Before using a container, wall effects should be excluded by thorough purging of the completely assembled device.

3.1.2
Gases Liquefied under Pressure

Gases liquefied under pressure are present in the container in two phases: a liquid phase, representing the major stock of substance and the gas phase above the liquid phase, formed by vaporized constituents of the mixture, depending on their boiling point (Figure 3.3 and Figure 3.4). The content in the cylinder is not correlated with the pressure, as the cylinder pressure is only dependent on the temperature of the liquefied gas (that is usually the temperature of the cylinder). High temperature means high pressure, low temperature means low pressure. If it cannot be ascertained that the mixture has been maintained constantly at the same temperature, it must be taken into account that the composition of gas phase is only dependent on the temperature of the liquid inside the cylinder: condensation of one or more components might take place resulting in the change of composition both of the liquid and the gas phase.

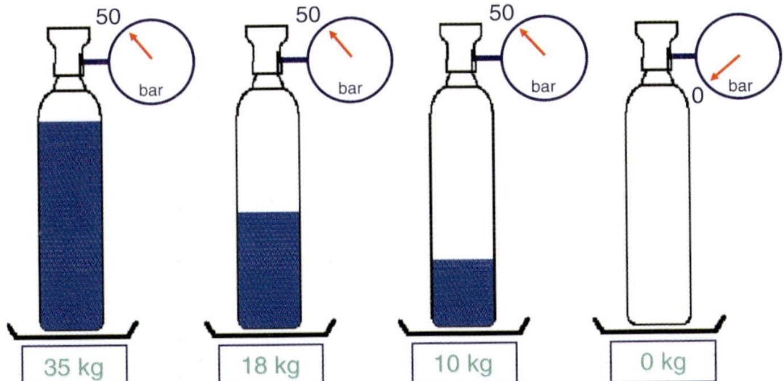

Figure 3.2 Gas cylinder with liquefied gas under pressure and different filling rates [55].

All components in the cylinder being in a dynamic equilibrium including condensation and vaporization during changing temperatures of the cylinder and thus the liquid in the cylinder (Figure 3.4). Moreover, when tapping gas phase the dynamic equilibrium of the cylinder will change permanently: pressure drop in this system leads to cooling down of the liquid, while the components with lower boiling points are irreversible lost by withdrawing the gas phase. It is a matter of fact that especially the low boiling inert gases for example in carbon dioxide or nitrous oxide would be affected: if a greater amount of gas phase is lost during taking the sample, 'good' results are pretended, but the contaminant air gases have been already vented by venting the gas phase.

To estimate the constituents of the liquid phase, one has to separate a well-defined portion of the liquid phase, which has to be completely vaporized (e.g., by heating of the sampling unit) before it enters the detector of the instrument.

In general, we can collect at least 5 essentials for a good and reproducible sampling of liquefied gases without altering the concentrations in the cylinder:

a) Before sampling, the gas must be allowed to reach ambient temperature. Usually, the cylinders are kept for 24 h at room temperature,

b) the cylinder to be analyzed must be homogenized again. To homogenize the constituents, the cylinder is rolled for an appropriate length of time to get samples with stable composition.
c) when drawing the sample only liquid phase has to be withdrawn, to avoid subsequent distillation of the lighter components.
d) as the sampler usually is of room temperature, any vaporization of liquid phase during the sampling process should be avoided.
e) in the sampling space a portion of liquid is collected. It is therefoe mandatory to secure the integrity of this space by a bursting disc or safety valve.

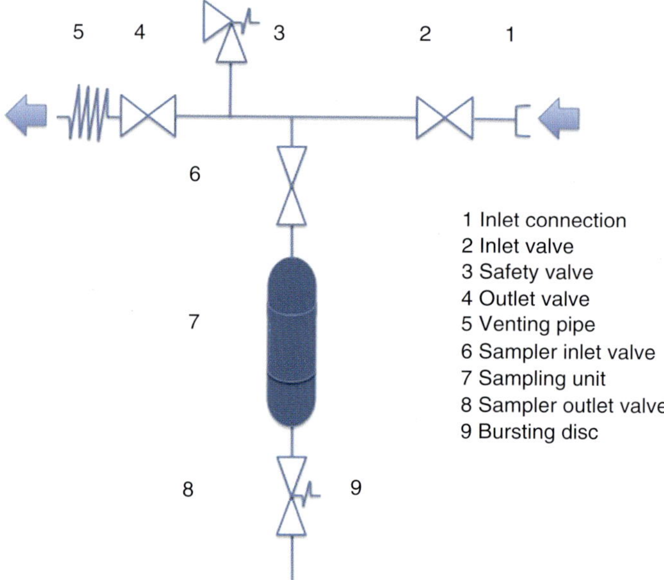

Figure 3.3 Typical easily built sampling unit used both for liquid and permanent gases [54].

The composition of gas and liquid phase might be different, depending on the impurities or the composition of the sample. A representative check of unknown gaseous samples is possible only by analyzing both gas and liquid phases. In case of medicinal gases, the samples are produced from a well-known and defined source, so it might be appropriate to analyze just one phase, depending on the later use of the gas if the distribution of the constituents is known for specific temperatures.

Since the gaseous phase is generally of lower density than the liquid phase, in an upright standing cylinder, the gas phase is present at the valve and can be tapped with the same precautions as for the permanent gases (see Figure 3.4), for example, by means of the right pressure reducer.

Sampling from the liquid phase is a bit more complicated: the concentration of constituents or impurities might differ in the gas or in the liquid phase.

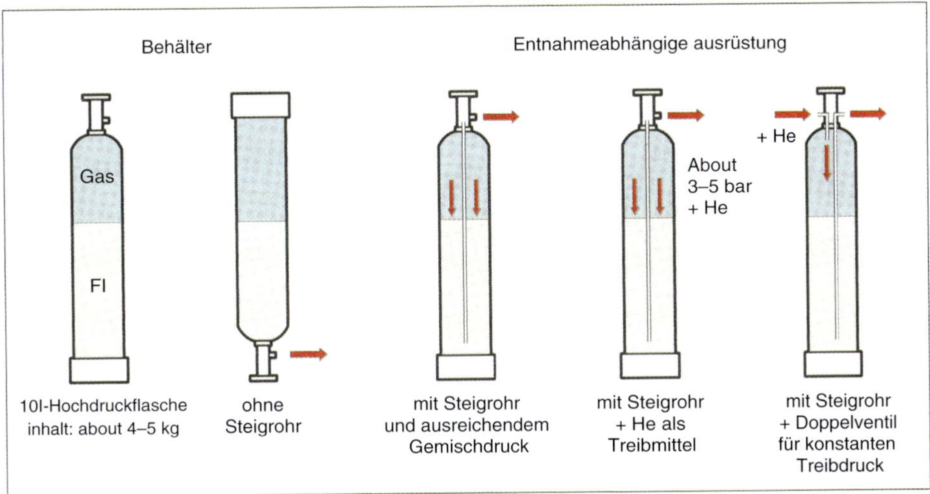

Figure 3.4 Tapping of liquid phase from a cylinder by turning it upside down, or by using a dip tube, or by using a dip tube and helium pressure [24].

For some gases, the cylinders may have a dip tube (see Figure 3.4) allowing direct access to the liquid phase, without turning the cylinder upside down.

All pressure-regulating devices have to be removed as they remain without function when entered by liquid phase. From the uninterrupted, pressurized flow of the liquid phase, a portion has to be sampled by appropriate valves (e.g., gas chromatographic systems often have suited valves (3.2) which cut off a defined portion of the liquid phase, releasing it into the low-pressure sampling line). This portion of liquid has to be vaporized completely before it enters the analytical instrumentation, which is a very sophisticated process. Even if small parts of the original sample remain liquid or if lighter components are vaporized before the main part of the sample enters the analyzer, a faulty analysis would result. Typically, a heated sampling line would provide complete vaporizing of all constituents of a sample by the time it reaches the probe.

3.1.3
Cryogenic Gases

Sampling of cryogenic gases puts the highest requirements for sample taking: as the cryogenic liquid vaporizes it releases the different constituents in the sequence of their boiling points. To separate a representative portion of the liquid it is necessary that the sampling equipment is cooled down to the temperature of the liquid. Only if this is provided an aliquote part can be separated from the reservoir. Small portions of cryogenic liquid can develop big volumes of gas, it is important to safely cut off a volume of the liquid and to vaporize this liquid completely.

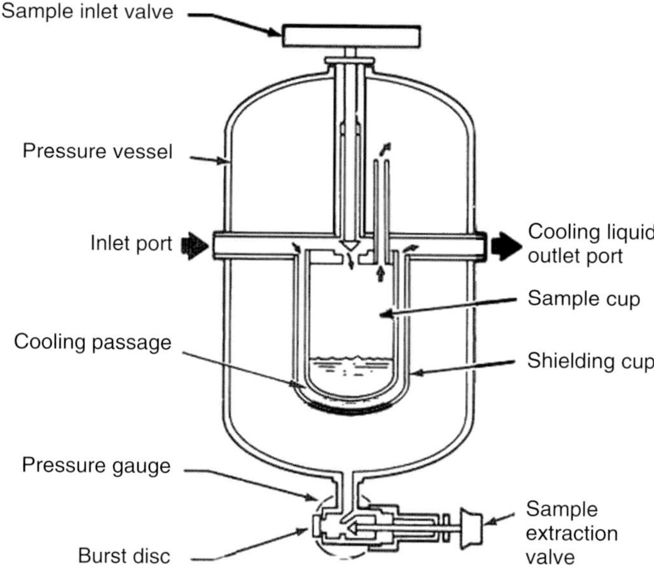

Figure 3.5 Schematic Flow of a cryogenic sampler (Cosmodyne).

Additionally the sampler material and all connecting hoses have to be carefully chosen to withstand the cryogenic temperatures and the developed pressure after vaporizing.

To achieve that, professional sample-taking equipment has been developed by the gas industry [55]. Very common are the device supplied by Linde and by Cosmodyne Corp.

Figure 3.5 shows the flow schematic of a Cosmodyne-sampler. At the inlet (fat arrow on the left hand side the source of cryogenic liquid is connected to the sampler, the outlet (fat arrow on the right hand side) is kept open.

Sample inlet valve (on top) and sample extraction (outlet) valve (on bottom) are kept closed.

Opening the liquid reservoir (mobile tank, tank, or tanker) provides a steady flow of cryogenic liquid through the device, cooling down the sample cup. The complete cooled down is reached; when on the outlet a continuous flow of cryogenic liquid leaves the sampler. This flow must not be directed to any object in the vicinity.

Only when the device has been cooled down sufficiently, the sample inlet valve on top of the unit is carefully opened to allow the cryogenic liquid entering the sample cup, after one minute it is closed again. The valves at the liquid reservoir have to be closed now, too.

The cryogenic liquid caught in the cup now vaporizes in the the space of the sampler, finally reaching a typical pressure of about 27,5 bar (400 psig).

After complete warming up of the sampler by opening the sample extraction valve the sample can be withdrawn and analyzed.

3.2 Gas Analytical Methods

Developed from so-called Hempel burettes (Figure 3.6), the early gas analytical methods seemed to be more handcraft rather than sophisticated analyses. Supported by suitable absorption media, one component after the other was extracted from the gas, finally ending up with a residual gas bubble that could not be absorbed under ambient conditions.

Hempel burettes are used for the determination of oxygen (assay) containing copper coils in an absorption solution. If shaken with a portion of oxygen, oxygen is absorbed in the solution, leaving a small gas bubble not reacting with the absorption solution, consisting mainly of argon and nitrogen. The gas bubble is collected in the graduated part of the burette top, where a reading allows to determine the assay of oxygen being >99.0 vol% [56].

While in the early years of gas analysis, the methods widely used were volumetric methods, we can now rely on a number of well-developed and versatile methods to analyze gases or the impurities. Generally, most of the common analytical methods are also suited for the analysis of medicinal gas. Restrictions exist only by the small amounts of substance passing through the probe; or when the analyte has to be processed in a certain way, for instance, solving of gaseous components in water will be successful only with gases that dissolve to a high extent in water, such as ammonia or hydrogen chloride. In the field of medicinal gases, only carbon dioxide dissolves to a remarkable extent in water, the other gases (oxygen,

Figure 3.6 Hempel burette [56].

nitrogen, nitrous oxide, helium) are not so soluble in water; from the newer gases also, nitric oxide, which is quite reactive with air and also is very reactive with water, forms nitric acid.

Analytical methods for gases must thus be adapted to the small amounts of substance carried by the sampling line often with high velocity to the probe. Depending on the properties of the impurities and the carrier gas, it might be sufficient just to use a minimum of different analyzers, especially when considering the basic assumption that, if the gas is of known source, the common impurities, and constituents are well known.

The choice of gas analytical methods in the European Pharmacopoeia comprises simple processes with reliable instruments, which have proven their reliability over the last 30 years (Table 3.1). Looking back to the discussion in the beginning, to yield a reliable analysis of medicinal gases, one would need some essential instruments, depending on the nature of the gas to be analyzed and the frequency of the analyses. In general, the industry uses continuous analytical methods and noncontinuous analytical methods for their checks and controls during manufacturing.

Continuous methods (co, continuous sampling) are those methods where the gaseous sample is transferred to a probe in a continuous, uninterrupted flow either in a bypass or directly with the main flow of the gas to be analyzed. Instrumentation for this kind of methods sometimes requires larger amounts of gas to provide a continuous flow of the analyte through the probe and to reach appropriate detection limits. It could happen that sometimes the need of flow exceeds the amounts of gas that can be provided by cylinders, as is the case for the trace analysis of some high purity gases. Fortunately, the specification limits for medicinal gases do not reach those red lines, so all the continuous methods described in the Pharmacopoeia can also be used by the common gas lab analyzing a cylinder.

Table 3.1 Gas analytical methods as described in the Pharm. Eur. [57].

Section in this book	Titles	Method	Reference in Ph. Eur.
3.2.1	IR spectrometry	co	2.2.24
3.2.2 under 3.2.1	Gas chromatography	ss	2.2.28
under 3.2.1	Carbon dioxide in gases	co	2.5.24
	Carbon monoxide in gases	co	2.5.25
3.2.3	Nitric oxide and nitric dioxide in gases	co	2.5.26
3.2.4	Oxygen in gases	co	2.5.27
3.2.5	Water in gases	co	2.5.28
3.2.6	Sulfur dioxide	co	2.5.29
3.2.7	Oxidizing substances	ss	2.5.30
	Nitrous oxide in gases	co	2.5.35
3.2.8	Test tubes	ss	2.1.06

ss = single sampling.
co = continuous sampling.

Methods using single sampling (ss) are typically gas chromatography and some "antique" wet chemical determinations of oxidizing agents. The test for oxidizing agents was in the beginning part of the monograph for nitrous oxide and carbon dioxide. Gas chromatography requires taking a discrete sample to be processed and the specific contents can be isolated from each other by a column and be measured afterward. In the gases field, this kind of analysis require a thorough isolation and processing of the sample; technically speaking, one needs the sample loops, valving, and piping, which complicate the analysis and the design of the piping (containing fittings and connections which are often source of cumbersome analytical deviations). The technical difficulties usually can be overcome by a well-trained lab staff.

3.2.1
Infrared (IR-) Spectrometry

Spectrometry based on spectra in the region of 2.5–15.4 μm is generally called *infrared spectrometry* (IR) [56]. Although it is a widespread analytical method in industry and science, IR spectrometry of gases is special: to achieve a satisfying resolution, extraordinary path lengths for the cells are needed. Typical path lengths are 10 or 20 m, received by mirrors, allowing the beam to pass several times (thus prolonging the way through the medium) through the cell.

The infrared radiation is generated by an emission source, passes through a monochromator and the substance, and finally a detector detects residual light. Many of the air gas constituents in low concentrations consist only of one atom (such as the rare gases) or of two atoms of the same kind (such as the major components in ambient air, nitrogen, and oxygen, totaling up to 99 vol%).

To generate visibility in infrared spectrometry, at least two different atoms connected with atomic bonding in a molecule are needed, which is the case for most of the anthropogenic air contaminants: typical examples are carbon dioxide and carbon monoxide, which give strong and sharp absorption bands, while gases which are not IR-active, such as oxygen or nitrogen or the rare gases, helium, neon, argon, xenon cannot be detected by IR spectroscopy.

The intensity of the measured infrared absorptions of gases suffers usually from the low density of the gases, requiring long pathways of the sample cells to obtain a measurable effect. More information on the fundamentals of IR-spectroscopy can be found in [58].

For the analysis of known impurities or constituents, the measurement advantageously covers only those parts of the IR spectrum where the strongest or the characteristic absorptions are measured. The so-called nondispersive infrared instruments (NDIRs) for purity control (here the impurities of the gases are measured) are very widely used (Figure 3.7), e.g. for carbon monoxide and carbon dioxide.

The selective sensitivity of the detector is achieved by an appropriate absorption generated by a reference cell filled with the gas to be analyzed. The pathway or

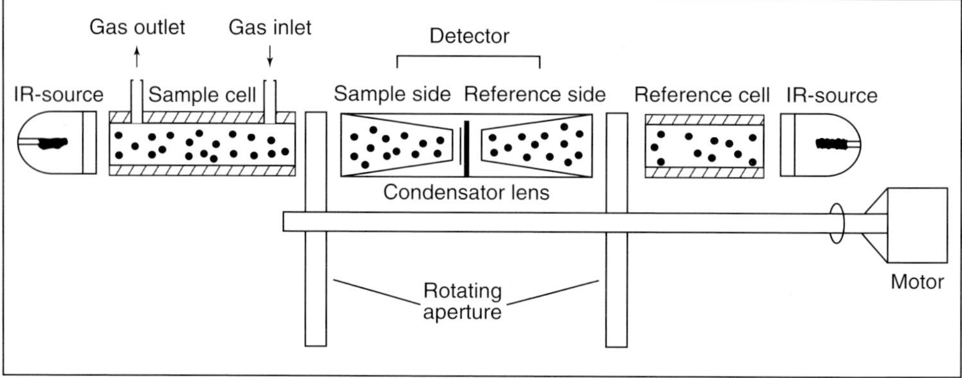

Figure 3.7 Principle of NDIR spectroscopy [59].

concentration of the gas in the reference cell is adapted to the concentration range of the intended measurement [59].

In case of the measurement of the assay of infrared active gases, such as carbon dioxide, carbon monoxide, and nitrous oxide, the concentration range to which the instrument is adjusted usually covers only the region between 95% and 100%, thus reaching satisfying precision. Although different instruments can measure the same component, they cannot be exchanged, as the concentration ranges are different.

Instrumentation used in the analysis of medicinal gases usually is designed in a way that the detector cells are sensitive only to the component to be analyzed and furthermore the detector cell is also adapted to the expected concentration of the impurity in the gas.

The intensity of absorption in IR spectroscopy is (besides the strength of the absorption itself, caused by the strength and masses of the involved atoms) dependent on the number of molecules in the pathway of the absorption cell. Two difficulties are inherent: different components can absorb at similar wavelength, so they cannot be distinguished or, when one strong absorbing component is much stronger than the others, as for the carrier gas, other absorptions can be hidden by the strong absorption. In this case, alternative analytical methods have to be chosen.

3.2.1.1 **Calibration**

Depending on possible effects of other impurities, the instrument needs calibration either with single sample calibration gases (containing only the one component to be analyzed) or with calibration gases simulating the live composition of the gas to be analyzed with changing amounts of the analyte in order to catch possible cross sensitivity.

The reference gases used depend on the sort of analysis to be performed (see Table 3.2).

Table 3.2 Reference gases as described in the Ph. Eur.

No	Gas	Analyte (to be determined)	Method	Reference gas (a)	Reference gas (b)
1	Argon	Identity and impurities	GC	5 ppm CH_4 R1 / 5 ppm N_2 R1 / 5 ppm O_2 R in Ar R1	—
2	Air, medicinal	CO_2	NDIR	<1 ppm CO_2/21 vol% O_2 R / 79 vol% N_2 R1	500 ppm CO_2 R1 / 21 vol% O_2 R/ 79 vol% N_2 R1
		CO	NDIR	<1 ppm CO/21 vol% O_2 R / 79 vol% N_2 R1	500 ppm CO R1 / 21 vol% O_2 R / 79 vol% N_2 R1
		SO_2	FA	21 vol% O_2 R / 79 vol% N_2 R1	0.5–2 ppm SO_2 R1 / 21 vol% O_2 R / 79 vol% N_2 R1
		NO/NO_2	CLD	<0.05 ppm NO/NO_2/21 vol% O2 R / 79 vol% N2 R1	2 ppm NO R / N_2 R1
3	Carbon dioxide	NO	CLD	CO_2 R1	2 ppm NO/CO_2 R1 or in N_2 R1
		Total sulfur	FA	CO_2 R1	0.5–2 ppm H_2S R1 / CO_2 R1
4	Carbon monoxide	CO	NDIR	CO R	5 vol% N_2 R1 in 95.0 vol% CO R
		CO_2	GC	300 ppm CO_2 R1 in CO R	—
		CH_4	GC	100 ppm CH_4 R in CO R	—
		H_2	GC	300 ppm H_2 R in CO_2	—
5	Helium	Assay	GC	He R	—
		CH_4	NDIR	He R	50 ppm CH_4 R/He R
6	Nitrogen	Assay	GC	Ambient air	Nitrogen R1
		CO_2	NDIR	Nitrogen R1	300 ppm CO_2 R1 in N_2 R1
		CO	NDIR	Nitrogen R1	5 •ppm CO R in N_2 R1
7	Nitrogen, low oxygen	CO_2, CO, CH_4, Ar + O_2, H_2	GC	Ambient air	Nitrogen R1

(*continued overleaf*)

Table 3.2 Reference Gases (Continued).

	Nitric oxide	CO_2	GC	3000 ppm CO_2 R1 in N_2 R	—
		N_2	GC	3000 ppm N_2 R in He R	—
		NO_2	UV	Nitrogen R1	400 ppm NO_2 in N_2 R1
		N_2O	GC	3000 ppm N_2O R in N_2 R	—
8	Nitrous oxide	Assay	NDIR	N_2O R	5 vol% N_2 R1 / 95 vol% N_2O R
		CO_2	GC	300 ppm CO_2 R1 / N_2O R	—
		CO	GC	5 ppm CO R/N_2O R	—
		NO/NO_2	CLD	N_2O R	2 ppm NO R/N_2 R1
9	Oxygen	CO_2	NDIR	O_2 R	300 ppm CO_2 R1 / N_2 R1
		CO	NDIR	O_2 R	5 ppm CO R/N_2 R1
10	Oxygen 93%	CO_2	NDIR	O_2 R	300 ppm CO_2 R1 / 7 vol% N_2 R1 / 93 vol% O2 R
		CO	NDIR	O_2 R	5 ppm CO R/N_2 R1
		NO/NO_2	CLD	<0.05 ppm NO + NO_2/21 vol% O2 R / 79 vol% N_2 R1	2 ppm NO_2 R / N_2 R1
		SO_2	UV	7 vol% N_2 R1 / 93 vol% O_2 R	0.5–2 ppm SO_2 R1 / 93 vol% O_2 R/7 vol% N_2 R1

Gases with reference R, R1 are defined in the Ph. Eur. under "Chemicals."

The right choice of composition and concentration of the needed calibration gases are defined in the monographs that are used for the definition of necessary measurements.

All NDIR methods are well suited to perform continuous measurements, for example, during production of the gases.

3.2.2
Gas Chromatography

Fifty years of development of this method generated a great number of excellent books and descriptions [60, 61], so we are able to just briefly describe the general conditions of gas chromatographic methods and focus on the specialties connected with the analyses of gases.

The benefits of GC methods are obvious: the analysis of more than one constituent of the analyzed gas in one analytical run is possible, with only a few restrictions set by the detector and the separation columns used. For analyses of gases in industrial environments, the GC methods are as well suited as IR analysis; they are not continuously working methods: discrete samples enter the analytical system and are analyzed one after the other. Depending on the time requirement to elute all components of a mixture, the time lag until the next analysis might take at least a few minutes. Complications can occur, either by summing up of components with long retention times (back-flush techniques have to be applied) or by unintended addition of components to the next chromatogram, especially those with higher molecular weight and long retention times (Figure 3.8) or when one peak covers the other(s).

As gases, especially those for medical use, are of low molecular weight and of nearly similar physical properties, there is neither a complicated injection system nor are sophisticated separation techniques necessary to analyze the given product mixtures coming out of either air separation plants or from chemical synthesis. Moreover, the small amount of substance carried with a sample volume of gas works best with packed-type columns instead of capillary columns. Figure 3.10 shows a typical view in a GC-lab as in the beginning 1980s.

In fact, the sensitivity of the detector is one of the reasons why mixtures of air gases cannot be analyzed in one turn: while the inert gases and methane are well separated by molecular sieve-type adsorbents, hydrocarbons require a different detector (FID, flame ionization detector) as well as a synthetic adsorbent material for efficient separation.

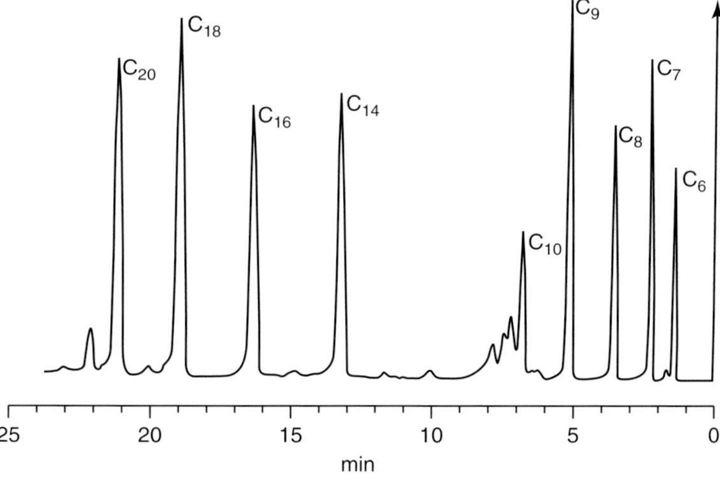

Figure 3.8 Typical chromatogram (hydrocarbons separated with a Porapak® Column) [62].

Numerous authors have worked to simplify the conditions and constructions of instruments suited for both types of instrumentation. A good combination is usually a dual-channel instrument, equipped with a THC (thermal conductivity detector) with molecular sieve column for the separation of inert gases and methane, as well as an FID with a column for the separation of hydrocarbons, and so on. To take into account the difference in sensitivities of the different detectors, the instrument should be equipped with two injection systems. "Methanizers" are used especially for the trace analysis of carbon oxides (carbon monoxide and carbon dioxide). These are little catalyst boxes switched in the flow to reduce the carbon oxides to methane. The two carbon oxides, which have already passed a molecular sieve column earlier, thus pass one after the other into the catalyst, and are reduced to methane, generating two peaks of methane, the area of which is related to the original amounts of carbon monoxide and carbon dioxide (Table 3.3).

To avoid any corruption of the samples directed to the instrument, gases are usually switched pneumatically into the port of the detector. Typically these switches work with a defined space purged first and filled with the gas to be analyzed. Switching while the gas is still flowing would generate slight pressure differences in the switch, thus also a different quantity of gas finally purged into the detector (Figure 3.9). The picture showing on the right hand side the switch as a circle with the different six connections. The coil between connection No. 1 and the connection No. 4 is the so called sample loop: the coil has a fixed volume which is charged with sample gas before switching into the detector.

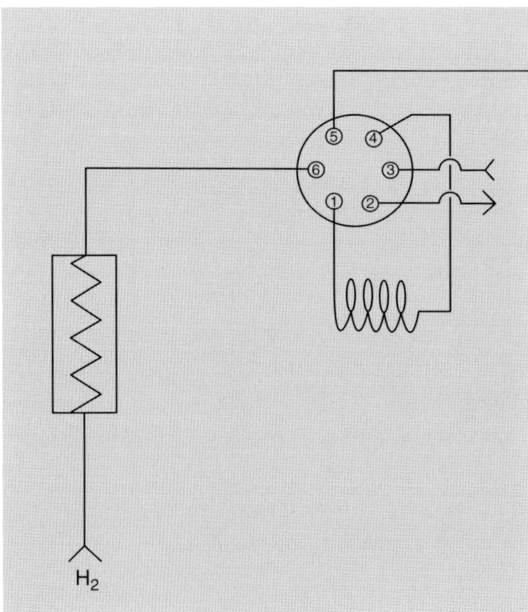

Figure 3.9 Typical GC switch port (right).

Table 3.3 List of components, appropriate detectors, and columns for separation (Ph. Eur.).

1) Air gas mixtures composed of O_2, Ar, N_2, CO

Column: stainless steel, size: 2 m, diameter 3 mm
Stationary phase: molecular sieve for chromatography (150–180 μm, pore size 0.5 nm)
Temperature: column 50 °C/thermal heat conductivity (THC) detector 150 °C
Carrier gas: helium
Flow rate: 10 ml min^{-1}
Injection, for example, 25 μl

2) Hydrocarbon mixtures CH_4, C_2H_6, C_3H_8, C_4H_{10}, and isomers until C5

Column: stainless steel, size: 2 m, diameter 4 mm
Stationary phase: ethylvinylbenzene–divinylbenzene copolymer (177–250 μm)
Temperature: column 95 °C/injection port and flame ionization detector(FID) 240 °C
Carrier gas: nitrogen, helium
Flow rate: 10 ml min^{-1}
Injection, for example, 1 ml

3) Carbon dioxide, carbon monoxide

Column: stainless steel, size: 2 m, diameter 4 mm
Stationary phase: molecular sieve for chromatography (150–180 μm, pore size 0.5 nm)
Temperature: column 50 °C/injection port and flame ionization detector(FID) 130 °C
Detection: FID plus methanizer
Carrier gas: helium
Flow rate: 60 ml min^{-1}
Injection: loop injector

4) Nitrous oxide in nitric oxide

Column: stainless steel, size: 3,5 m, diameter 2 mm
Stationary phase: molecular sieve for chromatography (150–180 μm, pore size 0.5 nm)
Temperature: column 50 °C/thermal heat conductivity (THC) detector
Detection: THC
Carrier gas: helium
Flow rate: 15 ml min^{-1}
Injection: loop injector

The switch has two different positions: 'purging' connects 2 and 3, 4 and 5, and 6 and 1; and 'charging' connects 1 and 2, 3 and 4, and 5 and 6.

To avoid pressure effects, it is recommended to proceed as follows: after sufficient purging time, the flow of gas is *stopped* and the switch is agitated by air pressure, to purge a portion of gas in the sampling loop into the detector. The continuous flow of carrier gas now transports the "bubble" of sample gas through the pipe to the column, where the constituents are separated and finally they enter, one after the other, into the detector, generating a signal with an area proportional to the concentration of the component in the gas (Figure 3.8).

3.2.2.1 Calibration

For quantitative measurement, the use of calibration gases is recommended for each of the components to be determined. A calibration gas is a pressurized mixture supplied by a gas manufacturer with a certified analytical value and given precision of the determination, thus acting as a standard to calibrate the reading or the signal of the instrument in relation to the concentration of the same component. Depending on the composition of the gas to be analyzed, either single-component calibration gases or complete gas mixtures are recommended.

For necessary reference gases refer to Table 3.2.

Figure 3.10 GC-Lab in the eighties of the last century (Air Liquide).

3.2.3 Chemiluminescence

Both nitric oxides, nitric monoxide and nitric dioxide, are determined via a chemiluminescence method [62]. The method is based on the reaction of nitric oxide with ozone, resulting in excited molecules of nitric dioxide. When slipping back to normal state, a light radiation which can be detected with a detector operating at 1.2 μm wavelength is emitted. A photomultiplier tube enhances the signal, which can be recorded. The method is already exhaustively described in numerous publications; the fundamentals of the measurement can be found in [63] (Figure 3.11)

As usually both nitric oxides are contained in oxygen-containing gases, the method must be slightly altered to receive either the total amount of both the gases or the values of the concentration of each of the nitric oxides. Nitric dioxide

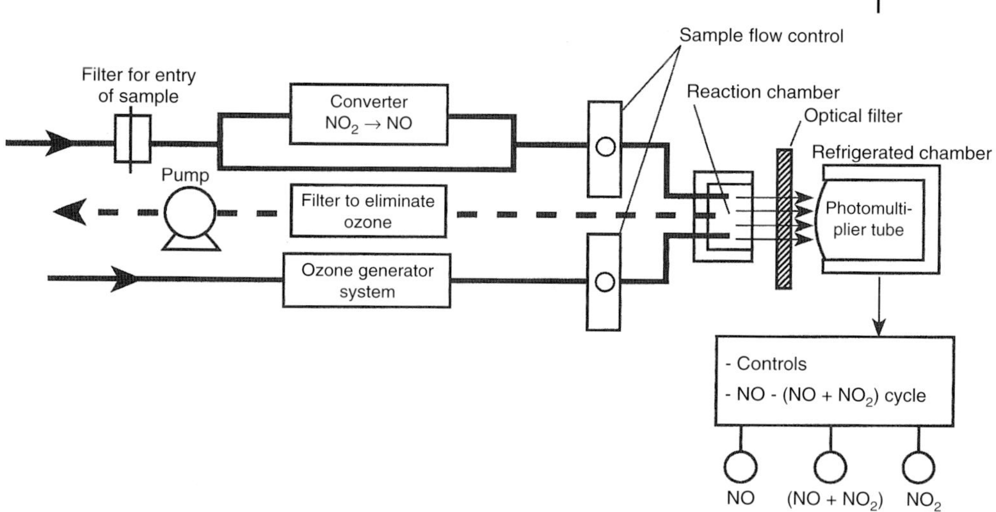

Figure 3.11 Principle of chemiluminescence spectroscopy [62].

in the sample gas will not enter in the excited state and thus would not give a luminescence radiation; it has to be reduced to nitric oxide first.

To reduce existing nitric dioxide, a converter is used. This converter reduces nitrogen dioxide to nitrogen monoxide leading to an indication of the sum of both components as response of the detector. To determine just one of the two components, the two values are collected, one with converter in the flow, the other without converter in the flow, representing the sum of nitric oxides ($NO + NO_2$) and nitric oxide (NO) alone, respectively. The subtraction of the concentration of NO from the total value yields the value for NO_2 in the original sample gas.

The ozone generator can be operated either with dried ambient air or with pure oxygen. Because of the possibility of organic impurities present in ambient air (e.g., in a lab) reacting with ozone, high purity oxygen from a cylinder is recommended for use.

3.2.3.1 **Calibration**

For calibration, the use of NO mixtures in the appropriate concentration range is the common method of choice. The efficiency of the converter has to be verified periodically with an appropriate NO_2 mixture. Regular calibration should include the verification of both the ozone generator and the converter. If both oxides should be determined with high precision, both components should be calibrated in their appropriate concentration range.

For necessary reference gases refer to Table 3.2.

3.2.4
Paramagnetic Measurement

The principle and general buildup of the instruments are well described in the literature, so we focus here on the manner in which to handle the instrument in the gas analytical environment [62].

Oxygen, because of two unpaired electrons in the outer shell of electrons, is one of the few elements in nature that is paramagnetic. This unique property gave cause for the early development of specific instruments, using the effect in a magnetic field for the detection of oxygen. As the effect depends on the amount of oxygen in the probe, the analyst has to correct for temperature and pressure if the instrument is not electronically compensated. The method is very easy to handle, with continuously working instruments following this principle being widely used in industry (Figure 3.12). In a commonly realized set-up the sample gas is purged through a glass foil, which is covered on one side by a solenoid. If the gas is paramagnetic it is attracted by the magnetic field. The generated flow is detected via a Wheatstone bridging in the middle of the coil.

3.2.4.1 Calibration

For the analysis of medical gases, there are two different layouts of the instrument: one for measuring the oxygen content of ambient or medical air, the measuring range lying between 20 and 23 vol% and the other one, to measure the purity

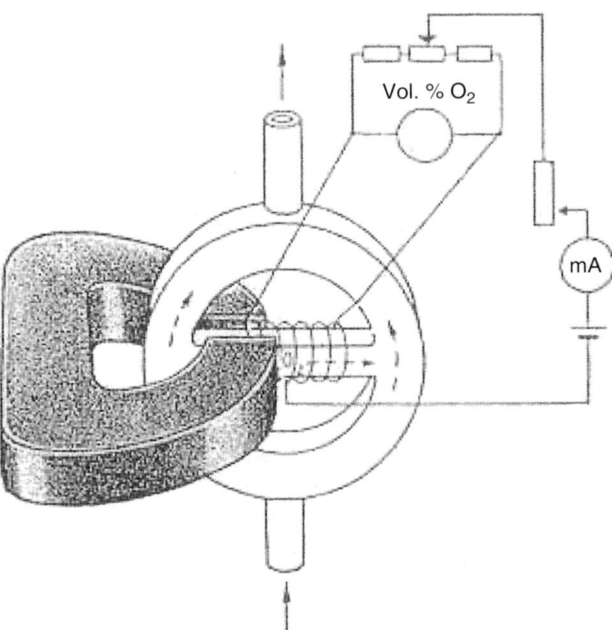

Figure 3.12 Principle of a paramagnetic detector [66].

or assay of pure oxygen. For that purpose, the instrument is equipped with an expanded measuring scale between 90 and 100 vol% of oxygen. This expansion allows for determining the high concentration of oxygen as crude oxygen, typically coming from the air separation plant (ASU).

Depending on the range in which the instrument is operated, either ambient air (20.9 vol% of oxygen) or calibration gases (pure oxygen with traces of nitrogen) have to be used to calibrate the instrument. Typically, one of the calibration gases is highly pure oxygen (minimum content of oxygen 99.995 vol%), while the other one is slightly contaminated oxygen (e.g., 99.0 vol% or as described in the Ph. Eur. European Pharmacopoeia [64]) with a zero gas such as nitrogen. Always available is ambient air, containing constantly with 20.09 vol% of oxygen which can be used as an easily available calibration gas for a functional test of the instrument and for the calibration.

3.2.5 Moisture Measurement

Moisture measurement is often performed either by dew-point measurement (a mirror is cooled down until moisture condenses on the surface [65]; the changed reflection of the mirror can be detected with an optical measurement) or with the help of hygroscopic substances, such as phosphorus pentoxide (Figure 3.13). The hygroscopic substance is precipitated as a thin surface layer on different materials. By absorbing water from the gas, the physical properties of the substrate will change: oscillation (piezoelectric), electrical capacity, or conductivity.

The most prominent measurement is the measurement of relative humidity in ambient air, but the trace determination of moisture in highly purified gases covers a complete different range of concentration. Oxygen or nitrous oxide vaporized from a cryogenic liquid are very dry compared to ambient air. Before administering they have to be enriched with moisture to prevent irritation of the sensible mucosa in the breathing area. Medical gases for inhalation usually are saturated with moisture before administering the gas to a patient to avoid any irritation of the patient's mucous membranes. Drying of gases before compression has practical reasons, as moisture would condense as liquid when the dew point is reached during or after compression of the gas.

To avoid liquid in pressurized cylinders, the moisture content must be below the dew point of the gas at cylinder pressure and at possible ambient temperatures (e.g., in middle Europe down to 0 °C) and at the pressure of the gas in the cylinder (usually 200 bar). At different pressures (e.g., in the piping system of a hospital), the pressure stays lower (typically at 10–15 bar); here higher moisture contents could be tolerable if condensation can be excluded.

Liquid in both cylinders and piping is not tolerated as it might cause corrosion and also facilitate possible settlement of bacteria.

When using moisture meters, the range of allowed concentrations for cryogenically produced gases seems rather high, the value of 67 ppm required by Ph.

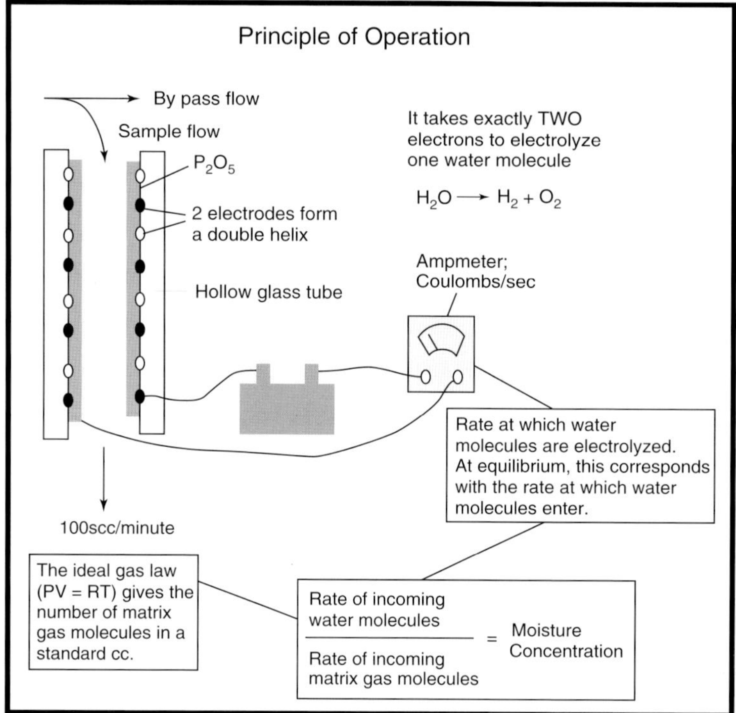

Figure 3.13 Principle of a P_2O_5 H_2O detector[1].

Eur. is usually out of scope of industrial gases. As the gases produced in an ASU are always vaporized from a liquid phase at cryogenic temperatures (−186 °C for oxygen, −196 °C for nitrogen), the gases from chemical sources (carbon dioxide, nitrous oxide) are very thoroughly dried before compression to high pressures (200 bar) to avoid any difficulties; keeping within the limits of the Pharmacopoeia is not of risk. On the other hand, for the compression of medical air for piping, the pressure range is quite low (below 15 bar) and so higher amounts of humidity are tolerated as there is no danger of condensation.

With the very dry gases from ASUs, some widely used alternative measuring principles have a tendency to age probes owing to the very low moisture content of the measured gas in continuous operation. This could lead, after a long period of uninterrupted operating time to erroneous values, indicating very low concentration of moisture in the gas, caused by insensitivity of the detector.

To avoid those effects, Ph. Eur. has defined the exclusive use of instruments based on P_2O_5 measuring cells.

As the active layer of diphosphorus pentoxide is transformed during use to phosphoric acid, the cell needs re-formation at regular intervals, which reduces the risk of corrupted cells to zero.

1) Meeco-Tigeroptics, Warrington, PA, USA.

3.2.5.1 Calibration

By absorption of water, phosphoric acid is formed. Since it is an electrical conductor, this changes the electrical properties of the cell. Two platinum wires are fixed in the layer, acting as electrodes when a voltage is applied across them. If water is absorbed by the layer, the measurable current follows Faraday's law in direct relation to the absorbed amount of water [65].

3.2.6 Fluorescence Analysis

Sulfur dioxide responds by emitting fluorescent radiation when excited with an ultraviolet radiation. For the excitement, when a wavelength of 210 nm is used, sulfur dioxide emits at a wavelength of 350 nm. The instrument thus consists of a source of ultraviolet radiation at 210 nm and a detector (photomultiplier tube) sensitive at the wavelength of 350 nm. To generate a stable fluorescence, the nitrogen concentration in the sample must not exceed the concentration of nitrogen in ambient air. A too high content of nitrogen would result in a notable quenching effect of the signal and thus lead to erroneous results [68] (Figure 3.14).

For the analysis of hydrogen sulfide and related compounds, the same instrument is used. For the analysis of sulfur dioxide, the instrument can directly be used. If sulfidic components are present in the sample gas, it must be passed through an oxidizing converter, to make sure that all sulfur components have been transferred to sulfur dioxide, which can be detected. By comparing the received

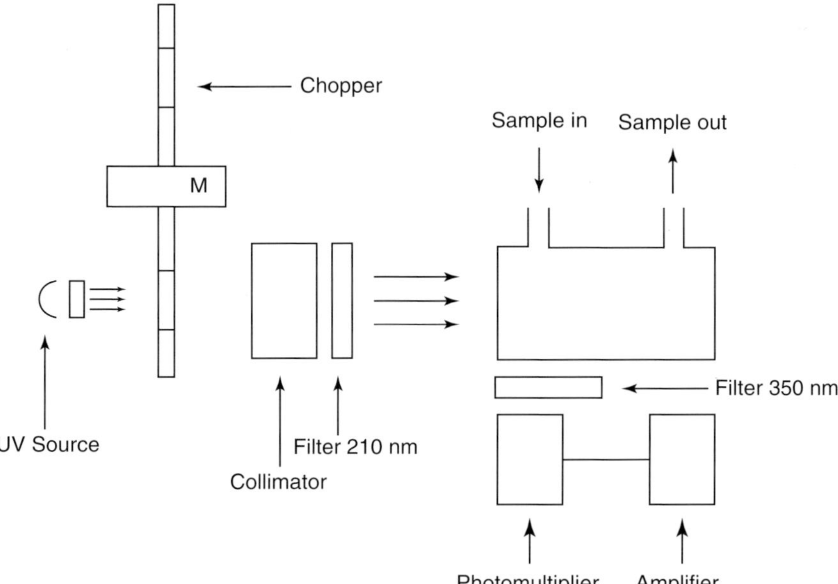

Figure 3.14 Schematic Flow-sheet of a fluorescence analyzer [68].

results with and without converter, the concentration of both contaminants can be determined separately.

3.2.6.1 Calibration

The instrument is usually calibrated with sulfur-free synthetic air, serving as zero gas and a calibration gas containing up to 2 ppm (V/V) sulfur dioxide. To avoid quenching effects which would adulterate the received result, calibration gases should have the same nitrogen matrix as the sample gas.

For necessary reference gases refer to Table 3.2.

3.2.7 Test Tubes

Test tubes are easy-to-use analyzers with selective sensitivity: Until the early 1980s, test tubes were the most commonly used method to analyze medicinal gases [69]. The lack of versatile methods led to the use simple methods. Unfortunately, test tubes are difficult to validate and produce a lot of debris often containing dangerous chemicals, acids, or caustic agents well mixed with pieces of broken glass, which are difficult to dispose of within the valid regulations.

As a consequence, the use of test tubes has decreased over the last 20 years. They are still in use, when checks are considered to be necessary outside of production, for example, for testing breathing air according to EN 12021 or at regular intervals to check gas outlets of medicinal gas pipeline systems (MGPS) according to EN ISO 7693-1. Test tubes are available for numerous components, some being relevant for medicinal gases, and they are always advantageous for rapid checks of ambient atmospheres for irritating or toxic components.

Generally, test tubes contain one or more chemicals reacting reliably with the constituents to be analyzed in the gas. This reaction is chosen very thoroughly to result in a color change clearly visible from the outside of the glass tube, progressively advancing with the amount of the constituent.

Originally following a US patent of Lamb and Hoover (see Figure 3.15) in 1919, Dräger developed the method to a kind of quantitative analysis, allowing a quick analysis of atmospheres, especially in coal mining.

While the original patent describes mainly a qualitative analysis, the present day tubes are capable of delivering trustful values of most of the common impurities in the air.

Although test tubes are easy to handle, it is useful to respect some fundamental advice to avoid erroneous results. Breaking the two ends of the tube opens the detector tubes and then they are inserted into a tube carrier. This tube carrier is adapted to the intended use; it can carry just one test tube. A common test tube carrier used exclusively for the analysis of medicinal gases can carry up to six test tubes through which the gas is passed for analysis.

As test tubes require different amounts of gas to reach the required sensitivity, the multiple carriers include some orifices for specific tubes to adjust the gas flow to receive the required quantity. This leads to predefined positions of the tubes,

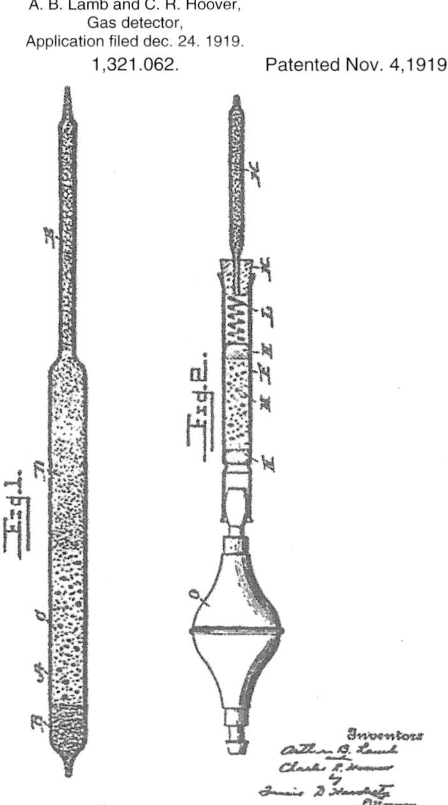

Figure 3.15 Drawing from the patent of Lamb and Hoover [70].

which cannot be changed (e.g., the SO_2 test tube must always be positioned at the position marked for it and must not be used in a position where CO is to be measured).

The chemicals in the test tubes can consist of one or more layers, depending on the reactivity of the analyte. Figure 3.16 shows examples for single-layer and double-layer test tubes. It is obvious that the gas flow corresponds with the indicated direction marked on the test tubes.

The gas to be analyzed is introduced into a tube either by a manually operated pump or by the pressure in the sampling line (which sometimes is needed to be reduced). In each case, the flow direction indicated on the outside of the test tube has to be maintained. This is especially important as the chemicals in the tube react with the constituent in a predefined way. A reversed flow direction would result either in no reaction, a deferred reaction, or at least in a nonquantitative reaction.

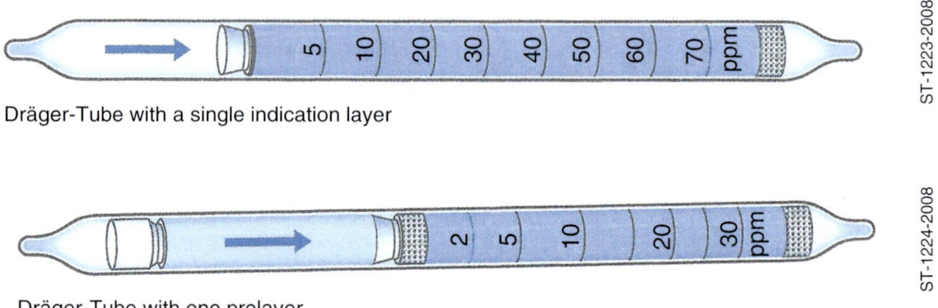

Figure 3.16 Examples for test tubes: single-layer and multiple-layer test tubes [70].

In more sophisticated tubes, a chemical is kept in a small glass container in the middle of the tube. Before usage, the tube is bent, the inner glass container is broken and the chemical can react immediately after release (Figure 3.17).

Exceptional care must be taken for measurement of constituents also available in ambient air (e.g., carbon dioxide, carbon monoxide, moisture). In this case, during the sampling procedure, one has to prevent even contact with ambient air (saturated with moisture far above the 65 ppm to be measured in the gas) containing carbon dioxide and carbon monoxide (depending on the environment) from entering into the test tube.

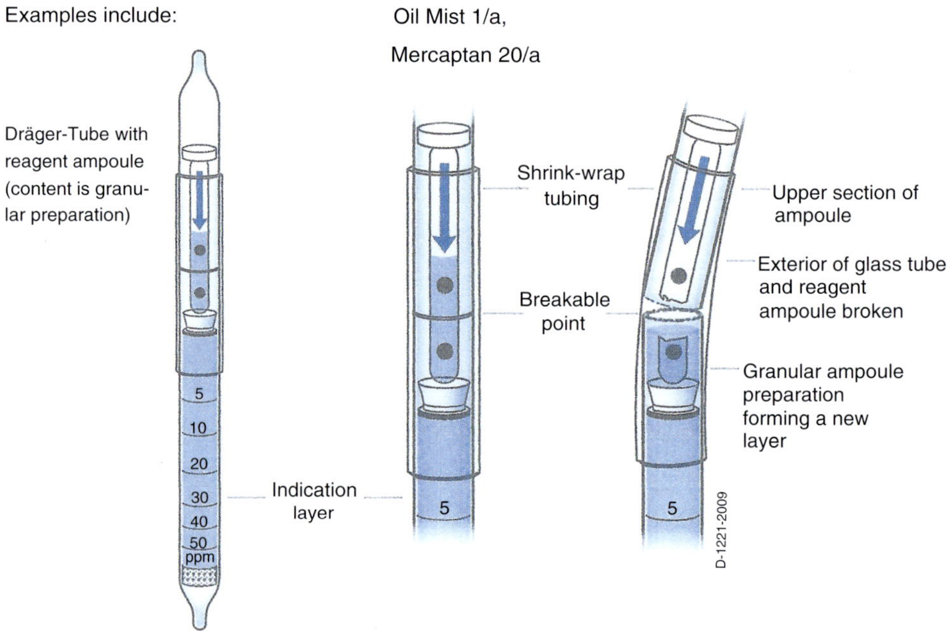

Figure 3.17 Dräger tube with ampoule inside [70].

In this case, all parts of the sampling line have to be purged with the gas to be measured and the sampling must be done under the flowing gas. All the instructions of the manufacturer should be followed. In general, test tubes work correctly only when the indicated temperature interval for the operation is maintained and, in some cases, the gas must not exceed or fall below the specified values in the instruction manual.

Given below are the descriptions of the reaction as described by the manufacturer (Figure 3.18) [70]:

- Test tube for carbon dioxide (10 strokes)
  The white indicator layer changes color to blue-violet when carbon dioxide is present in the sample gas. The tube uses the color change of methyl violet to give an indication of carbon dioxide.
- Test tube for carbon monoxide (10 strokes)
  The tube is filled with a visible pre-layer (orange) and an indicator layer (white). Passing gas with traces of carbon dioxide through the tube the indicator layer turns to a greenish brown by reduction of di-iodine pentoxide to elementary iodine.
- Test tube for hydrogen sulfide (10 strokes)
  The white material inside the tube changes color to pale brown, when hydrogen sulfide is present in the sample gas. Cross sensitivity to sulfur dioxide only at higher concentrations, not relevant for medical gases.
- Test tube for moisture (3 strokes)
  The indicator layer turns color from yellow to blue. With water in the sample gas a blue complex is formed. Mind the precautions to measure water as indicated above.

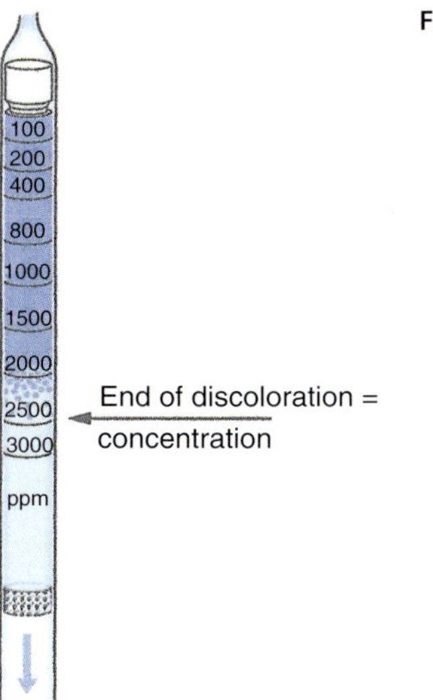

Figure 3.18 Example for coloring of a tube [70].

- Test tube for nickel tetra-carbonyl (20 strokes)
  The nickel tetra-carbonyl is cracked with iodine and the released nickel forms a red dimethyl-glyoxime complex. To achieve a good reaction, the tube is first purged by 20 strokes with the samples gas, then an internal ampoule with the reagent is broken and the nickel complex is formed. Is iron-pentacarbonyl present in the gas, only the carbonyl is decomposed and a brownish color is formed.
- Test tube for nitric oxide and nitric dioxide (5 strokes)
  A pre-layer oxidizes nitric oxide to nitric dioxide, the indicator layer showing up the total of the nitric oxides. The color of the indicator layer changes from pale gray to bluish gray.
- Test tube for oil (100 strokes)
  As the number of necessary strokes show clearly, the low amounts of present oil require a large amount of sample gas. Often, the analysis for oil from sampling containers with a test tube remains insufficient as the gas runs out. The reaction is based on the decomposition of oil (vapor and mist) by sulfuric acid. This works well only for hydrocarbon oil, but not for synthetic or silicon oils. For hydrocarbon oils, different sensitivities have also to be tested out before. Before starting the measurement, the inner ampoule containing the sulfuric acid needs to be broken.
- Test tube for sulfur dioxide (100 strokes)
  A pre-layer (gray) and an indicator layer (blue) react with present sulfur dioxide to form a white layer. There is no measurement when hydrogen sulfide is present.

Instead of using the hand-operated pump and counting the strokes to draw a defined volume through the tube, specified tubes can also used under flowing gas. The flow has to be adjusted to be constant (e.g., by an appropriate orifice), so that the time measurement delivers the amount of gas passing through the tube.

This kind of approach is facilitating tests when more than one component has to be checked with different tubes.

4
Monographs for Gases in the European and National Pharmacopoeias

4.1
European Pharmacopoeia Specifications

4.1.1
General Composition of Monographs

Among the numerous standards and specifications for technical gases, the monographs in the different pharmacopoeias give the relevant frame for the mandatory quality for the medicinal use of gases. As a comparison between the three pacemaking Pharmacopoeias – the US American, the Japanese, and the European Pharmacopoeia – shows that although the view on quality is equal, the means are still different [71] in some limiting values as well as in the analytical methods.

In the monographs of the substances, the permitted limits of impurities (contaminations) and the level of purity of the active ingredient (assay) are described. In the same monograph, a methodical description is given also of how these impurities or how these assays should be determined. By definition, Pharmacopoeia methods are regarded as being validated, so in fact the Pharmacopoeias are a collection of good and reliable information about nearly every common component to be used in a medicinal surrounding. The methods described can be replaced by methods with proven suitability to obtain the same result. In the common understanding, the suitability of an analytical determination method can be proved by a validation of the alternative procedure.

Although the European Pharmacopoeia is one of the oldest projects in the European environment, the stakeholders developed over the years. Once started as a project of the Council of Europe in 1964, the whole organization has been adapted to the growing needs of an increasing number of contract states during the last 50 years to become the European Directorate for the Quality of Medicines and Healthcare, based in Strasbourg, among other European and UN authorities.[1] The EDQM is today a directorate of the Council of Europe, one of its four departments is the European Pharmacopoeia Department, which is responsible for the European Pharmacopoeia Commission. The first edition of

1) Strasbourg is the domicile of the numerous European Authorities; besides the Pharmacopée, the European Court, the Human Rights Court, and the European Parliament hold their meetings there.

Medical Gases: Production, Applications and Safety, First Edition. Hartwig Müller.

the European Pharmacopoeia (Ph. Eur.) was released in 1964 and the current one is edition 8, the release of which started in January 2014. The Publisher of the Ph. Eur. now being the EDQM, the European Directorate for the Quality of Medicines and Healthcare in Strasbourg.

From the early beginning until now it has been important that the members of the treaty on the European Pharmacopoeia were not restricted to the members of the Council of Europe or to the European Union. At present, the Ph. Eur. is adapted in 47 signatory countries (among them 28 EU member countries) plus a number of organizations (such as the WHO and the EU). The directives about human and veterinary medicinal products (2001/82/EC, 2001/83/EC, and 2003/63/EC) form the legal basis for the European Union.

From the start, the monographs were written by numerous groups of experts, mainly being pharmacists or appropriate experts of related professions, for example, chemists or pharmacologists, who are familiar with the specific properties of the compounds to be described. As a result of the heritage of the different stakeholders, the monographs are mirrors of the development of the authors of the different articles. In these expert groups, usually led by a member of the European Pharmacopoeia Commission, the experts from the industry and/or from the relevant application discuss the mandatory qualities or the requirements toward the specifications in regard to the technical availability.

Monographs are developed by the expert commissions after initial steps by the members of the Pharmacopoeia Commission, usually the health authority of a country. The number of the gas-related Pharmacopoeia monographs has risen from 3 in 1984 to 12 in 2014 [72]. This is quite a remarkable development in the release of monographs, considering the amount of work that goes into each monograph before it is ready for release.

From the beginning, a major point of discussion was the intended use of the released monographs: on the one hand they should be the basis of industrial production lines, but on the other, they should be simple enough to be used by the simple pharmacist practitioner as well.

This issue also reveals contradictory targeting: industrial methods suited to produce and to maintain quality of large series of unique pharmaceutical products are completely different from lab methods used by the pharmacist, who, in his laboratory, is obliged to maintain the quality of his own piece-by-piece production and to check ready-made drugs in the pharmacy.

The intended preparation and analysis of medicinal gases requires special knowledge both of preparation and of handling of cylinders. Additional problems arise with the thorough sampling and analysis of gases; this is a field where the ordinary pharmacist reaches a limit if not specially equipped and educated. Furthermore, handling and storage of medicinal gases needs, for reasons of safety, special precautions for storage (see Chapter 7).

The Pharmacopoeia Commission, being well aware of this difficulty thus asked for a separation of the monographs into two parts: the PRODUCTION part and the TESTS part. Older monographs with direct roots in the beginning of the work on the European Pharmacopoeia (E.P.) still show this segregation, namely those for the classical medicinal gases oxygen, nitrous oxide, and carbon dioxide. In later releases of the monographs, the Pharmacopoeia Commission lost this methodology; in some of the monographs, all analytical methodology is kept under "PRODUCTION," while in others, the analytical methods appear exclusively under "TESTS" (see Table 4.1).

Another point affects the production of the compound, which is often the active ingredient in the later produced drug. Since the method of isolation or of synthesis of a substance is decisive for the achieved purity, namely, the concert of impurities in an active compound, the definition of the manufacturing method is decisive for the checks to be made to confirm pharmaceutical purity.

Not in all of the monographs is the method of manufacturing sufficiently fixed to specify a series of impurities to be checked at various occasions (e.g., during and after manufacturing, during or after transfilling, or during and after transport). This point being of utmost importance for active ingredients synthesized over many complicated steps from sophisticated starting materials, but is only of minor interest for gases, as the most important medicinal gases (oxygen and medical air) are both simple molecules or mixtures directly isolated from the ambient air after a purification operation.

Although easy to achieve and simple in their molecules, medicinal gases are directly incorporated into the vital systems of a patient; thus the products used for medicinal purposes do not tolerate uncertainties in the choice and the application of analytical checks.

Another aspect to consider for the gases is, compared to other states of matter, the different way of possible contamination, as all of them have to be handled under positive pressure (that means higher pressure than ambient atmospheric pressure). This principle results in a different spectrum of possible contamination risks compared to the handling of liquids or solids.

4.1.2
Use of Monographs in the Industry

Pharmacopoeia monographs are the main tool to develop industrial methods suitable for the production of medicinal gases. As the methods in the monographs are considered to be validated and also to be the recommended ones, validation in the industrial plants is reduced to the validation of the realized setup of instrumentation, namely, the so-called analytical racks.

Typically, these racks consist of the required instruments, appropriate piping, and a computerized control of all flows, pressures, and switching of the lines. For reasons of safety, pressures are restricted to the allowable pressure of the instruments. Usually, the design of the analytical rack allows both for analysis of the gas flow entering the cylinders before filling, as well as the analysis of a single cylinder, as stated in some of the monographs.

In the environment of an ordinary filling plant for medicinal gases, the analytical rack helps to make a reliable documentation of each batch of production and it thus controls the later release of the cylinders. The same instrumentation

Table 4.1 Monographs for medicinal gases in the European Pharmacopoeia (excerpt of the essentials).

Name Reference	Argon 07/2010:2407	Medicinal Air 01/2009:1238	Air Synthetic Medicinal 01/2008:1684	Carbon Dioxide 01/2008:0375	Carbon Monoxide 01/2011:2408	Helium 01/2008:2155
Chemical Properties / CLP-Classification [a]	Asphyxiant, GHS 04	Pressurized gas, GHS 04	Pressurized gas, GHS 04	Pressurized and liquefied gas, GHS 04	Asphyxiant, Toxic, Flammable, GHS 06, GHS 02, GHS 08, GHS04	Asphyxiant, GHS 04
Formula	Ar	$N_2 + O_2$	$N_2 + O_2$	CO_2	CO	He
Odour	not specified	not specified	odourless	not specified	not specified	not specified
Physical Properties/ Physical State	gas	gas mixture	gas mixture	gas, under pressure liquefied	gas	gas
Definition	Gas obtained by fractional distillation of ambient air.	Compressed ambient air	Mixture of nitrogen (1247) and oxygen (0417)	Content minimum: ≥99,5 per cent (V/V) of CO_2 in the gaseous phase	Gas obtained by steam reforming (catalytic oxidation) of hydrocarbons. Content minimum 99,5 per cent V/V of CO	this monograph applies to Helium obtained by separation from natural gas and intended for medicinal use.
	Content: min 99.006 per cent V/V of Ar, calculated by deduction of the sum of impurities found when performing the test for impurities and the water content.		Content: 95.0 per cent to 105.0 per cent of the nominal value which is between 21.0 per cent V/V to 22.5 per cent V/V of oxygen (O_2)			
Characters / Appearance	colourless gas	colourless gas	colourless and odourless gas	colourless gas	colourless, flammable gas	colourless inert gas
Solubility	at 20°C and at a pressure of 101 kPa, 1 volume dissolves in about 29 volumes of water	at 20°C and at a pressure of 101 kPa, 1 volume dissolves in about 50 volumes of water	at a temperature of 20°C and a pressure of 101 kPa 1 volume dissolves in about 50 volumes of water	at 20°C and a pressure of 101 kPa, 1 volume dissolves in about 1 volume of water	at 20°C and at a pressure of 101 kPa, 2,266 volumes dissolve in about 100 volumes of water	
Impurities	Specified: O_2, H_2O, detectable imp.: N_2, CH_4	CO_2, SO_2, NO, NO_2, oil, CO, H_2O	H_2O	NO, NO_2, CO, total S, H_2O	CO_2, CH_4, H_2, $Ni(CO)_4$, $Fe(CO)_5$, H_2O	CH_4, O_2, H_2O

a) The dangerous goods classification is not part of the Ph.Eur.

Table 4.1 (*Continued*)

Name **Reference**	**Methane** **01/2015:2413**	**Nitrogen** **01/2008:1247**	**Nitrogen, Low Oxygen** **01/2008:1685**	**Nitric oxide** **01/2008:1550**	**Nitrous Oxide** **01/2008:0416**	**Medical Oxygen** **01/2010:0417**	**Oxygen 93%** **04/2011:2455**
Chemical Properties / CLP-Classification[a)]	Flammable GHS 02, GHS 04	Asphyxiant, GHS 04	Asphyxiant, GHS 04	Toxic, Oxidizing, GHS 06, GHS 03, GHS 05, GHS 04	Oxidizing, GHS 03, GHS 04	Oxidizing, GHS 03, GHS 04	Oxidizing, GHS 03, GHS 04
Formula	CH_4	N_2	N_2	NO	N_2O	O_2	O_2 (+N_2 and Ar)
Odour	odourless	odourless	odourless	not specified	not specified	not specified	not specified
Physical Properties/ Physical State	gas	gas	gas	gas	gas, under pressure liquefied	gas	gas mixture
Definition	Content: minimum 99.5 Vol% of CH_4 This monograph applies to methane obtained from natural gas and intended for medicinal use	Content: minimum 99.5 Vol% of N_2	Content: minimum 99.5 per cent V/V of N_2. This monograph applies to nitrogen used for inerting finished medical products	Content: minimum 99,0 per cent V/V of NO	Content: minimum 98.0 % V/V of N_2O in the gaseous phase when sampled at 15°C	Oxygen contains not less than 99,5 Vol% of O_2	content: 90,0 per cent V/V to 96,0 per cent V/V of O_2, the remainder mainly consisting of argon and nitrogen. The monograph applies to medicinal not to domiciliary use.
Characters / Appearance	colourless gas, flammable in air	colourless, odourless gas	colourless, odourless gas	colourless gas which turns brown when exposed to air	colourless gas	colourless gas	colourless gas
Solubility	at 20°C and at a pressure of 101 kPa, 1 volume dissolves in about 27 volumes of water	at 20°C and at a pressure of 101 kPa, 1 volume dissolves in about 62 volumes of water and about 10 volumes of ethanol (96%)	at 20°C and at a pressure of 101 kPa, 1 volume dissolves in about 62 volumes of water and about 10 volumes of alcohol	at 20°C and at a pressure of 101 kPa, 1 volume dissolves in about 21 volumes of water	at 20°C and at a pressure of 101 kPa, 1 volume dissolves in about 1.5 volumes of water	at 20°C and at a pressure of 101 kPa, 1 volume dissolves in about 32 volumes of water	
Impurities	Specified: N_2, H_2O, Ethane, Propane, i-Butane (2-Methyl-propane), n-Butane	CO_2, CO, O_2, H_2O	O_2, Ar	CO_2, N_2, NO_2, N_2O, H_2O	CO_2, CO, NO, NO_2, H_2O	CO_2, CO, H_2O	CO_2, CO, SO_2, NO+NO_2, oil, H_2O

a) The dangerous goods classification is not part of the Ph.Eur.

is used for gases produced either in an air separation plant (ASU (air separation unit), namely oxygen) or by means of compressing ambient air after purification to produce medical air.

From the viewpoint of a larger scale of production, not all the analytical methods in the monographs are equally suited for an industrialization of the manufacture: gas chromatographic and moisture analyses often require a laboratory environment to be carried out in a sound and safe manner. A gas chromatographic analysis especially requires thorough knowledge of how to take samples and how to design the environment in the lab in respect of air conditioning, constant temperatures, and training of the lab personnel. This is a requirement not just for the chemically reactive gases (like nitric oxide) but also for common gases, such as nitrogen, argon, or carbon dioxide.

The Pharmacopoeia monographs here give valuable help in determining which reagents, which instrumentation, and which skill is needed to make a professional analysis of the medicinal gas.

4.1.3
Other Descriptions from the Ph. Eur.

Besides the substance-related monographs, the European Pharmacopoeia consists of general descriptions of analytical, purification, and separation methods for drug-related compounds. In the analytical methods chapter, some general analytical methods are described both for medical gases and, more generally, also for other substances [72].

The analytical methods specially assigned to medicinal gases were already described in detail in Chapter 3 of this book.

When reading the monographs, it is important to keep in mind, that the general chapters of the Pharmacopoeia in Volume I are integrated parts of every text later in the reference. Often the paragraphs about testing and storage, and thus determination of expiry dates, are indented with national (local) regulations, containing additional provisions.

Up to now, the European Pharmacopoeia does not contain specifications about reusable pressure vessels, which are used throughout the gas industry. As primary containment for the gases and the cryogenic liquid, this gap should be closed by EDQM in the near future.

One of the utmost constraints when applying European regulations and their correlated European standards for technical realization is the right of every member state to generate additional rulings, often complicating the community regulation in a way that national extensions form a local level of additional rules, only familiar to the local experts, thus requiring national appendices in the manufacturing procedures.

Reading and application of methods and monographs is, on the other hand, only possible with the knowledge and adherence of the general precautions and side conditions defined by the Ph. Eur. and by the supplementary local regulations.

Under the umbrella of the EU Directive 2001/83/EC, all drug-related substances to be administered to humans (and in the same way for medical devices as defined in 93/42/EC) have to be manufactured following the GMP Guideline, the guidelines for good manufacturing practice.

In recent years, GMP Guidelines have gained additional importance due to the globalization of both manufacturing and checking the production of drugs by national bodies. They provide a reliable framework of quality assurance rules for the manufacture of medical compounds thus adding valuable background for the industrial use of the Ph. Eur. and their monographs. To take into account the significant relevance of the GMP Guidelines, we will discuss the basic conditions in Chapter 5.

5
Production of Medical Gases — Special Handling to Comply with GMP Rulings

5.1
History – Gases Becoming Medicinal Products

Medical gases have been directly derived from technical gases, namely, from the application of pure oxygen. The use of oxygen for therapeutic objectives was obvious once the role of oxygen in breathing had been discovered. After the fundamental contributions of Priestley and Scheele in the 1870s [73] and the experiments of Lavoisier to distinguish between nitrogen and oxygen in air [74], in 1798, the first "Pneumatic Institution for inhalation gas therapy" in Bristol (UK) was established in 1799 [75, loc. cit. (2)].

It was not easy to develop application methods when the fundamental techniques to collect, store, and handle a gas had not yet been developed. Quality control was at that time preferably volumetric analysis based on absorbance of one component after the other by a chemical reaction. Administering oxygen to a patient suffered from insufficient devices, materials, and working techniques. Administering gas mixtures or gases other than oxygen was nearly impossible, owing to the lack of appliances such as respirators and mixing devices, and methods for analytical checks and control.

The improvement of the sourcing by invention of air liquefaction, the development of the application by further development of pumps, valves, back-flush restrictors, regulators, and, finally, the integrated respirator had a tremendous influence on administering oxygen, and later on a greater scale, to all other gas applications to humans.

On the other hand, the use of oxygen and other gases (mainly nitrous oxide) for medicinal purposes had begun early enough, well before the great development of synthetic drugs started in the beginning of the twentieth century. That was why oxygen and the gases were accepted drugs at the time that the pharmaceutical development was boosted by big inventions of the growing pharmaceutical industry. In the second half of the twentieth century, gas applications became very familiar to a generation of hospital doctors.

In the aftermath of some major drug disasters in different countries in Europe and America, the European Economic Community decided in the 1960s to release the drug directive 65/65/EEC [46] containing the essentials of reliable

Medical Gases: Production, Applications and Safety, First Edition. Hartwig Müller.

drug manufacturing. It was then that oxygen and other medicinal gases began to be accepted as drugs. Gases, directly entering into the body and instantaneously deploying their pharmacological impact, first gained considerable attention of health-care professionals, as evidenced by the addition of three monographs on oxygen, nitrous oxide, and carbon dioxide to the European Pharmacopoeia in the 1980s.

In the early 1990s, the Medical Directive was amended by the Medical Device Directive (MDD) 93/42/EEC [47] setting in force a segregation between drugs having a direct or indirect pharmacological impact on the functions of the human or animal and the medical device, where the physical properties are used to support functions of the body or support activities of the surgeon, thus affecting also medicinal gases such as carbon dioxide, sulfur hexafluoride, and octafluorocyclobutane, among others.

With the growth of the European Community and inclusion of most of the former EFTA countries and finally the establishment of the European Economic Area, since the early 1990s [76] the PIC, the Pharmaceutical Inspection Convention (originally established in 1970), has become a major driver for improvement of the pharmaceutical understanding and the development of necessary methods to control the manufacture of drugs for human use.

With integration into European legislation, the PIC/S (Pharmaceutical Inspection Cooperation Scheme) was created as a parallel expert panel to PIC, PIC and PIC/S work together in parallel on a worldwide basis with 40 member countries (2013) as an informal arrangement between Regulatory Authorities of the different countries.

Also on a worldwide basis, the "International Conference on Harmonization of Technical Requirements for Registration of Pharmaceuticals for Human Use" (ICH, see [77]) is bringing together the experts of the authorities of Europe, Japan, and the United States with experts from the pharmaceutical industry and providing the ICH guidelines for Efficacy, Multidisciplinary, Quality, Safety, and Consideration (EMA see [78]), which is another powerful driver of regulations and legislation.

All these efforts resulted in the pharmaceutical codex for human drugs as laid down in the directive 2001/83/EC [18], now released for the members and associated members of the European Union (EU) as the "Rules Governing Medicinal Products in the European Union." Here also, the *European Pharmacopoeia* was defined as the binding source for market authorization applications.

The Codex covers in 10 volumes the most important aspects of the manufacturing of drugs and contains the basic legislation and the framework of regulations and necessary side conditions, as indicated in the table that follows.

The next step in pharmaceutical development was the consolidation of the existing directives in the releases of 2001/82/EC [17] and 2001/83/EC [18], for veterinary and human drugs, respectively. The separation between active substance gas (as oxygen or nitrous oxide) and medicinal gas (medical air or the oxygen/nitrous oxide mixture) improved the clarity of the action of the authorities. In Eudralex and the other regulations we found necessary guidelines, also as list or check-lsit.

EudraLex – The Rules Governing Medicinal Products in the European Union[a].

Volume 1 Pharmaceutical Legislation: Medicinal Products for Human Use
Directive 2001/83/EC (consolidated 16 November 2012)
Volume 2 Pharmaceutical Legislation: Notice to Applicants and Regulatory Guidelines Medicinal Products for Human Use
2a Procedures for Marketing Authorization
2b Presentation and Content of the Dossier
2c Regulatory Guidelines
Volume 3 Scientific Guidelines for Medicinal Products for Human Use
Preparation of Marketing-Authorization Applications for Medicinal Products for Applicants: Quality, Biotechnology, Nonclinical, Clinical efficacy and Safety, Multidisciplinary
Volume 4 Good Manufacturing Practice (GMP) Guidelines
Directive 2009/94/EC for Human Use and Directive 91/412/EEC for Veterinary Medicinal Products
Part I Basic Requirements for Medicinal Products
Part II Basic Requirements for Active Substances Used as Starting Materials
Part III GMP-Related Documents: Site Master File, Q9 Quality Risk Management, Q10 Pharmaceutical Quality System, MRA Batch Certificate, Template for "Written Confirmation"
Annexes: Annex 6: Manufacture of Medicinal Gases; Annex 15: Qualification and Validation; Annex 16; Certification by a Qualified Person and Batch Release; Annex 17: Parametric release (just to name the most relevant of the annexes)
Volume 5 Pharmaceutical Legislation Medicinal Products for Veterinary Use (not referenced in this book)
Directive 2009/9/EC Amending 2001/82/EC Relating Medicinal Products for Veterinary Use
Volume 6 Notice to Applicants and Regulatory Guidelines for Medicinal Products for Veterinary use (not referenced in this book)
6a Procedures for Marketing Authorization
6b Presentation and Content of the Dossier
6c Regulatory Guidelines
Volume 7 Scientific Guidelines for Medicinal Products for Veterinary Use (not referenced in this book)
Quality, Safety and Residues, Efficacy, Immunological
Volume 8 Maximum Residue Limits Guidelines (MRL) (not referenced in this book)
Council Regulation (EEC) No. 2377/90
Volume 9 Guidelines for Pharmacovigilance for Medicinal Products for Human and Veterinary Use (not referenced in this book)
Directive 2012/26/EU Amending 2001/83/EU as Regards Pharmacovigilance, and Regulation (EC) 736/2004 (Establishment of a European Medicines Agency)
Volume 9B – Pharmacovigilance for Medicinal Products for Veterinary Use (not referenced in this book)
Volume 10 Guidelines for Clinical Trials (not referenced in this book)
Chapter I: Application and Application Form
Chapter II: Safety Reporting
Chapter III: Quality of the Investigational Medicinal Product
Chapter IV: Inspections
Chapter V: Additional Information
Chapter VI: Legislation

[a]Eudralex is the collection of the pharmaceutical legislations in the European Union, governing the medicinal products for human and veterinary use. It consists of 10 volumes.

These we tried to collect, wherever possible in tables, as we often stayed with verbal transfer to be sure not to drop important wording. Htis usually done without further identification, except the reference no.

It is not the objective of this book to give advice to applicants of market authorization, so we will not explain Volumes 1–3 in detail. We just propose to explain the principle and the contents of the “Good Manufacturing Practice” (GMP), by focusing on Volume 4 and the Annex 6, which give the details for daily aspects in the manufacturing of medicinal gases, relevant both for the manufacturer and for the inspecting authority.

Besides the proof of efficacy and the absence of adverse reactions, the uniform and reliable quality of a drug is the key feature of every drug brought to the market in Europe to be administered to humans.

Part I of the GMP Volume 4 describes thoroughly the basic requirements for the manufacturing of medicinal products. The core is the Pharmaceutical Quality System (PQS), which is needed to ensure that all conditions are followed and the final product is ready for the intended use. Often, the manufacturing of a drug is a multistep process, which is fixed by the market authority. The pharmaceutical manufacturer is obliged to use this process along the written actions and workflows as deposited at the authority.

All activities must be organized and documented in such a way that all changes that could influence the quality are collected and safely recorded. The PQS contains all basic requirements for the training of personnel, the appropriate premises and equipment, the documentation and all side activities connected to the production of the drug [79].

Important part of the PQS is the observation of the running production concerning the reached quality level. This continuous documentation is called the Product Quality Review (PQR) and collected in specified periods of times to be presented to the competent authority. The PQR helps to identify current difficulties in the manufacturing or in the supply chain. Part II of the GMP Volume 4 is dedicated to the manufacture of the starting materials, setting up here the basic requirements for using active ingredients as starting materials. This chapter will gain importance for the manufacturing of liquid oxygen, being doubtless the active ingredient of a drug. This part is the result of an amendment of the medical directive to regulate the manufacture of active ingredients or the Active Pharmaceutical Ingredients (APIs).

Annex 6 fixes the special conditions to be followed if the active substances or the active ingredients of the manufactured drugs are gases. It will be referenced under the corresponding headings in the general GMP.

5.2 Classification of Gases or Gas Mixtures as Medicinal Products

The approach of a drug being effective, only as hazardous as tolerable and of safe quality was fixed in the regulations and was worked out in a number of international working panels. Medicinal gases found their place in the early

1980s in the European Pharmacopoeia (Ph. Eur.) as a part of the efforts to create a common market for drugs and medicines. For the first time, uniform specifications for medicinal gases for the whole of Europe were agreed upon in the European Pharmacopoeia Commission (already in those days, the European Pharmacopoeia Commission was part of the Council of Europe, an organization whose members were not with the same as the member states of the European Economic Community).

Following the release of Directive 65/65/EEC and the appropriate amendments and consolidations (2001/83/EC and 2004/27/EC), the classification of medicinal gases as drugs came into effect in the different countries of the European Union. This was quite a time-consuming process: while the original member states and United Kingdom had already adopted the legislation in the 1970s and 1980s, the new members of the European Union are still in the process of adopting the EU directives with appropriate national legislation.

Generally, the medicinal gases oxygen and nitrous oxide have been accepted as drugs in most of the countries, with the pharmaceutical inspections being carried out with increasing intensity. The property of being regarded as a drug is defined in the directive by Article 2 [18]:

a) *any substance or combination of substances presented as having properties for treating or preventing disease in human being; or*
b) *any substance or combination of substances which may be used in or administered to human beings either with a view to restoring, correcting, or modifying physiological functions by exerting a pharmacological, immunological or metabolic action, or to making a medical diagnosis.*

In simple words, this means the legislator distinguishes between drugs as such (to defeat a specified disease, para 2a), functional drugs (to support specific functions of the body, para 2b), and drugs for diagnostic use. If at least one of these three conditions matches the intended application, then the medicinal product is to be classified as drug. Above all, the intention of the supplier on how to present the product on the market decides on whether the substance is a drug within the legal framework.

When the MDD came into force in 1993, all medical devices as defined in the Directive were excluded from the Drug Directive. To help to classify a compound or a gas as a drug or a medicinal product, one would have to refer to the borderline guidance of the Commission.

MEDDEV classifies gases for inhalation (therapy) as drugs, conforming to article 2b of the Directive and gas mixtures for *in vivo* diagnostics likewise; the components of the gas are seen as active substances, diluted in air (oxygen plus nitrogen q.s.).

On the other hand, carbon dioxide is a medical product when applied for dilatation of body caves. While the medicinal applications of carbon dioxide are difficult to verify via clinical trials, its use as a medical product for this indication clearly indicates the marketing as a medical device. Again the presentation in the market chosen by the supplier is in combination with the intended application, the relevant step for the decision about the nature of it being a drug, a medical device, a food, or others.

In case of gases, another distinction might become important: the active ingredient of the substance being that component in the gas which exerts in the latter drug "*a pharmacological, immunological, or metabolic action with a view to restoring, correcting, or modifying physiological functions or to make a medical diagnosis*". In a mixture, this is not often clear, as all the gases replace oxygen in the lung vessels, thus having a kind of physiological effect. In the sense of the Directive, they might be active ingredients. The deployment of medicinal understanding followed the development of the industry: in the late fifties and sixties great amount of steel cylinders were substituted by liquid delivery (Figures 5.1 and 5.2).

The production of liquid oxygen in an air separation plant can be regarded as the production of an active ingredient; Part II of the GMP has to be referenced. When a quantity of the liquid is separated to supply a hospital, the cryogenic liquid is dedicated for pharmaceutical use and the process of transfilling (e.g., in a tank truck) is in some countries to be admitted by the authorities with a production license.

Following the Directive and the GMP, the contents of the truck have to be analyzed according to Ph. Eur. and before delivery to a customer, the truck has to be released by a Qualified Person and documented in the Batch Release Register. As already mentioned, medical production is only a small part of the oxygen production of an Air Separation Unit (ASU), so in the daily practice, the Qualified Person is not all the time present in the ASU. Before the truck has reached the customer tank in the hospital, the driver should have received the release for the

Figure 5.1 Medical oxygen filling in steel cylinders (early 1950', Picture: Messer).

Figure 5.2 Truck for liquid transport of oxygen in the early 1960s (Messer).

contents of his transportable tank. The confirmation of the release is part of the documentation supplied during delivery.

With refilling the tank in the hospital, the pharmaceutical process at the supplier comes to an end: with handout of the metering of the quantity of liquid oxygen, (in case of validity of Annex 6, part 42, also handing out the results of the previously performed analysis) and confirmed by the responsible person of the hospital, the pharmaceutical process is finished. From this moment, the oxygen in the tank is the responsibility of the hospital and can be used to feed the gas supply system. If the tank is a customer tank, exclusively filled with dedicated transportable tanks and is used for the refilling of small transportable liquid tanks [meaning in case of validity of Annex 6, pt 42, at home or at a hospital] no analysis after the filling is necessary, if the delivered oxygen is accompanied by the analytical certificate of delivering tanker.

In the following table, for the most important of the different gases and gas mixtures their legal status in the countries of the European Union have been presented (Table 5.1).

5.2.1
Conclusion

Gases and gas mixtures have been used professionally in the health-care environment since the beginning of the twentieth century. With the beginning of the industrial air separation, the availability of oxygen was no longer a problem. Until the late 1950s, the supply of oxygen was managed with pressurized gas in steel cylinders and switched later to liquid (cryogenic) supply. Depending on their application, medicinal gases can be drugs or medical devices.

Table 5.1 Status of medicinal gases as medicinal products in the European Union (own source).

Gas	Application	Country								
		FR	IT	ES	PT	BE	DE	DK, SE, NO, SF	CH	UK
Pure gases										
Ar	Plasma coagulation	MD	No status	No status	No status	No status	No status	No status	No status	No status
Ar, He, N_2	Cryotherapy	No status	No status	No status	No status	No status	No status	No status	No status	No status
CO_2	Laparascopy	MD	MD	MD	MD	MD	MD	MD	MD	Drug
	Cryo dermatology	MD	No status	No status	No status	No status	No status	No status	No status	No status
	Cell culture	MD	MD	No status	No status	No status	No status	No status	No status	No status
N_2O	Anesthesiology	Drug	Drug	Drug	Drug	Drug	Drug	Drug	Drug	Drug
O_2/N_2	Medicinal air	—	—	—	—	—	Drug	—	—	—
O_2	Support of breathing	Drug	Drug	Drug	Drug	Drug	Drug	Drug	Drug	Drug
Mixtures										
N_2O/O_2	Anesthesiology	Drug	Drug	Drug	Drug	Drug	Drug	Drug	Drug	Drug
NO/N_2	Lung diseases	Drug	Drug	Drug	Drug	Drug	Drug	Drug	Drug	Drug
CO, He, O_2	Respiratory	No status	No status	No status	No status	No status	Drug	No status	No status	Drug
CO_2/O_2	Breathing support	No status	No status	No status	No status	No status	Drug	No status	No status	Drug
	Hearing loss	No status	No status	No status	No status	No status	Drug	No status	No status	Drug
SF_6, C_2F_8, C_3F_8	Ophthalmic mixtures	MD	No status	No status	No status	No status	No status	No status	No status	MD

Drug: Medicinal Products Directive 65/65/EEC (2001/83/EC).
MD: Medical Device Directive 93/42/EC (2007/47/EC).
No status: no status requested by local authority.

In some countries, gases were not in the focus of the authorities at the start of the new legislation, but this has been changing during the last 40 years. Table 5.1 reflects recent developments.

5.3 Basic Requirements (Volume 4, Part I) ([79] – GMP-Guidelines)

To secure pharmaceutical quality was one of the main objectives of the European directives concerning the medicinal products for human use, starting with 65/65/EEC, ending up in the recent 2001/83/EC. "The Rules Governing Medicinal Products in the European Union" contains guidance to interpret directives 91/356/EEC (amended by 2003/94/EC and 91/412/EEC).[1]

It is important to accept that not just the contents of Annex 6 are relevant for the manufacturing of medicinal gases but also the common understanding on how to produce medicinal products under pharmaceutical directives. Important parts of part I were revised last year and are now coming into force, such as the chapters about personnel (Chapter 2), quality control (Chapter 6) which will be analyzed below together with the other important chapters about documentation (Chapter 4) and self-inspections.

5.3.1 Pharmaceutical Quality System (PQS) and ICH Q10

PQS is the expression used for the pharmaceutical quality assurance system as mentioned in the directives (2003/94/EC and 91/412/EEC) to describe the obligations of the holder of a market authorization. It is meant as a mandatory prerequisite for a manufacturer to produce medicinal products [79].

Good manufacturing practice – basic requirements [79].

Manufacturing processes	Clearly defined, systematically reviewed, proven capability of consistently manufacturing
Critical steps	Identified and validated
	Necessary prerequisites for GMP are provided
Personnel	Qualified and trained
Premises and space	Adequate premises and space, suitable equipment and services, correct materials, containers, and labels, approved procedures and instructions, suitable storage and transport
Equipment and service	Suitable, adapted to the needs

(*continued overleaf*)

[1] Eudralex is the collection of the pharmaceutical legislations in the European Union, governing the medicinal products for human and veterinary use. It consists of 10 volumes.

(*Continued*)

Materials	Correct materials, checked before entering the manufacturing
Containers and labels	Proved and released
Storage and transport	Suitable and without compromising the quality of the product
Instruction and procedures	Written and clearly instructing, approved, and in accordance with PQS, the procedures are carried out correctly and operators are trained to do so
Records	Manually or by recording instruments: (i) to demonstrate the compliance of the process, (ii) to take note of significant deviations for later investigation and root cause analysis, and (iii) to allow for traceability of the whole batch
Distribution	Takes account of the Good Distribution Practice
Recall	Sustain a recall system from the market or supply system
Complaints	Complaint management system, procedure to examine and prevent reoccurrence

The explanation contains the general requirements for quality management systems (QMSs), as described for pharmaceutical companies in the EN ISO 13485 [80] or according to ICH Q9 (Risk management) [81] and ICH Q10 (Pharmaceutical Quality System) [82]. The quality manual is a description of the management system and the responsibilities of the management; the GMP is part of the PQS, ensuring that the products are consistently produced and tested following the market authorization.

It is the role of senior management to keep that system running by adequately sourcing, receiving, and commenting on reports on a regular basis. A detailed instruction on how to create a management system according EN ISO 13485 can be found in CEN ISO/TR 14969:2004 [83].

In addition, the ICH guideline Q10 on PQS [82] is included in the guidelines for further clarification and reference. The PQS in ICH Q10 is shown to work in an integrated approach with ICH Q8 (Pharmaceutical Development) and ICH Q9 (Quality Risk Management (QRM)) for industry and regulators. Quality is a product of science and knowledge and the PQS covers the whole life cycle of a product.

The PQS *elements* are as follows

- Monitoring systems for process performance and product quality
- Corrective Action/Preventive Action (CAPA)
- Change management system
- Management review.

The PQS *enablers* are as follows:

- Knowledge management
- Quality risk management.

For the common industry, the reference is EN ISO 9001, but this standard is not regarded as suitable for health-care companies, as the binding of the market authorization is not taken into account: if there are innovations or changes in relevant parts of the manufacturing process, a change management process has to be started, including the appropriate validations and additionally the change in the market authorization. Changes must not be introduced as long as they have not proved their efficacy; they have to be validated in view of the quality and the effects of the product on the patient.

The PQS is a wide-spanning management system, keeping in touch with all steps of production and defining measures to keep in order the process running, for example, self-inspections, error analysis, QRM, and finally ending up with the Qualified Person confirming that the products are manufactured in accordance with the existing manufacturing license and the documentation stored therein [82].

5.3.2 **Personnel**

Besides qualification and training, empowerment of the personnel plays a key role for setting a QMS into operation. The key role thus is not just at the operational level, it is equally to be demanded also in the management: production and quality departments often have a sensible relationship, when the management does not equally define the roles and limits of the two. On the other hand, the two departments could work hand in hand supporting each other with necessary resources and to complement each other. Such a sensible relationship requires the complete support and cooperation of the highest management of the pharmaceutical company.

Eudralex, Vol. 4, Chapter 2:

Head of Production Department (HoP)	Head of Quality Control (HoQ)
Appropriate storage and production of the products	approve or reject starting materials, packaging materials, intermediate, bulk and finished products
Improve instructions relating product operations	Ensure that all necessary testing is carried out and recorded
Ensure their strict implementation	approve specifications, sampling instructions, test methods, quality control procedures

(*Continued overleaf*)

Head of Production Department (HoP)	Head of Quality Control (HoQ)
Ensure production records are evaluated and signed	approve and monitor contract analysts
Ensure qualification and maintenance of his department, premises, and equipment	ensure the qualification and maintenance of his department, premises and equipment
Ensure that appropriate validations are done	ensure that the appropriate validations are done
Ensure that required initial and continuous training of the personnel is carried out	Ensure that required initial and continuous training of the personnel is carried out

Shared Responsibilities of HoP and HoQ
authorisation of written procedures
monitoring and control of the manufacturing environment
plant hygiene
process validation
training
approval and monitoring of suppliers of materials
approval and monitoring of contract manufacturers and other GMP related outsourced activities
design and monitoring of storage conditions
retention of records
monitoring of compliance with GMP
Inspection, investigation, and taking of samples, in order to monitor factors affecting quality
participation in management reviews of process performance, product quality, the QMS and advocating continuous improvement
ensuring that a timely and effective communication and escalation process exists to raise quality issues

In the table, we have used the listing from Chapter 2 of GMP Volume 4 [83], clearly indicating the different tasks of the two stakeholders as also the field of joint responsibilities. In daily practice, it helps a lot if each of the managers respects the different roles of the players on behalf of realization of the GMP-oriented QMS and keeping the QMS running.

5.3.3 Premises and Equipment

The existing guidelines of an average gas company to establish industrial gas plants do not naturally fit the needs for medicinal gas plants. During design of a medicinal

Figure 5.3 Preparation area in a medical filling plant.

gas plant, special care has to be taken to avoid later quality risks [84], high rising running costs for validation, and for cleaning and maintenance of the medicinal gas plant. Sometimes, existing industrial gas plants are refurbished and become medicinal gas plants, which might be critical in view of future production flow (segregation of manufacturing status) and cleaning aspects (separation of clean and not clean areas) (Figure 5.3).

Typically, industrial gas plants are situated in industrial areas often with relevant pollution, air and soil contaminations. In spite of the fact that the risk of contamination for gases is very low, moreover, the general impression for the inspector might not be the best, if paint is peeling and birds' nests are under the roof; debris collecting in the corners of the filling stands, or dust flaking from the roof construction.

Although the gases filled in cylinders do not have any contact with the environment of the filling plant (one exemption is medical air, produced by compression of ambient air at the site), the picture for the inspector remains unfriendly as the possibility that the outside of the cylinders can be contaminated must be taken into account. He is used to walk in clean rooms, meeting white-clothed people with mouth and hair masks and gloves, and here the gas plant displays itself quite different to the pharmaceutical world.

Even if the gas remains uncontaminated, in a rough environment other GMP rules remain violated and additionally the question of the outer surface of cylinders entering a hospital has increasing impact. GMP Annex 6 states under points 9 and 45 that filled gas cylinders should be delivered "to customers in clean state compatible with the environment in which they will be used" [85].

To avoid a situation, such as having to explain the inspector what the gases are, what purity they have, and where they can be contaminated, which leads in the audit situation rather quickly into unforeseen difficulties, one has to accept the inspector's view, facilities to produce drugs are clean, they have a cleaning plan posted, and the production is organized in a logic and simple way, avoiding confusion between empty cylinders, full cylinders, and released cylinders.

The premises should be large enough to be able to clearly separate all these different working areas, and a clear system of marking supports the workers at their job, which could be effectively done by a logic uninterrupted and continuous way of the cylinders from reception of the empties to final release of the full cylinders through the plant.

5.3.3.1 Annex 6

Fine-tuning is done for the medicinal gas plant in Annex 6: Handling of all vessels for medicinal use should be done in separated areas, except only those operations, which are performed also under GMP standards. The premises should provide sufficient space for separation of the different gases and the different stages of processing [85, pt. 7].

Exemplary Process stages (according to GMP Guideline, Annex 6)

AWAITING CHECKING
AWAITING FILLING
QUARANTINE
CERTIFIED
REJECTED
PREPARED DELIVERIES

5.3.4 Documentation

All procedures and instructions have to be approved as described in the management system, critical manufacturing steps have to be identified and validated, as in the case of significant changes of the processes [86]. All necessary recordings have to be made to demonstrate the full compliance with all described procedures.

Deviations are recorded and investigated to determine the root cause and to implement corrective actions. Complaints are examined in the same way to identify the root cause and to prevent recurrence.

Documentation is the key issue to demonstrate a running and living PQS. Documentation is essential to prove the elements of the PQS, as there are (among others):

- Trained personnel (processes, GMP as needed)
- Conduction of internal audits
- Generation and supply of controlled documents
- Validated processes
- Requirements reflected in SOPs
- SOPs along the practice exercised
- Batch release decisions
- Nonconformance to be investigated
- Complaints to be investigated
- Correction and prevention plans to be recorded
- Management reviews.

All activities which directly or indirectly influence the quality of the product have to be recorded, monitored, and checked with the objective to find immediately after occurrence, the root cause for errors or quality deviations.

From the viewpoint of good documentation practice, the two types of documentation, (i) directions and requirements and (ii) records and reports should be distinguished. Instruction documents should be written or at least in a form in which a printout can be easily produced.

GMP documentation:

- Site Master File (SMF)
- Instructions, such as
  - Specifications
  - Manufacturing Formulae, Processing, Packaging, Testing Instructions
  - Procedures (SOPs)
  - Protocols
  - Technical agreements (e.g., for external (outsourced) activities)
- Record/Report Type
  - Records (provide evidence of actions taken)
  - Certificate of analyses (provide a summary of testing results)
  - Reports (document of projects or investigations, together with results, conclusions, and recommendations).

Documents within the PQS should be regularly reviewed and the principles of good documentation practice should be applied:

- Handwritten entries should be made clear and indelible.
- Records should be made at the time of action and activities concerning the manufacture should be traceable.
- Alterations made in any document should be signed and dated.

Specific documents about manufacturing have to be retained. The local legislation here often gives the time period, at least 1 year after the expiry of a batch to which the record relates or 5 years after certification by the Qualified Person.

Specifications should be available for starting and packaging materials and for the finished products, the necessary subjects for which are given below [86, p. 4]:

Eudralex, Vol. 4, Chapter 4:

Specifications for starting materials
(a) A description of the materials, including – The designated name and the internal code reference – The reference, if any, to a pharmacopoeia monograph – The approved suppliers and, if reasonable, the original producer of the material – A specimen of printed materials (b) Directions for sampling and testing (c) Qualitative and quantitative requirements with acceptance limits (d) Storage conditions and precautions (e) The maximum period of storage before re-examination

Specifications for finished products
(a) The designated name of the product and the code reference where applicable (b) The formula (c) A description of the pharmaceutical form and package details (d) Directions for sampling and testing (e) The qualitative and quantitative requirements, with the acceptance limits (f) The storage conditions and any special handling precautions, where applicable (g) The shelf-life

The "documentation" chapter contains additional, complex descriptions about the composition of manufacturing formula and processing instructions, packaging instructions, and also the contents of the batch-processing record and the batch packaging record, as for the reception protocol. With respect to the length of this book, we have collected the keywords from Eudralex for each type of record in the table below:

Eudralex, Vol. 4, Chapter 4:

Manufacturing formula and processing instructions	**Packaging instructions**
The manufacturing formula should include (a) The name of the product, with a product reference code relating to its specification	(a) Name of the product; including the batch number of bulk and finished product (b) Description of its pharmaceutical form, and strength where applicable

(b) A description of the pharmaceutical form, strength of the product, and batch size
(c) A list of all starting materials to be used, with the amount of each, described; mention should be made of any substance that may disappear in the course of processing
(d) A statement of the expected final yield with the acceptable limits, and of relevant intermediate yields, where applicable
The manufacturing formula should include

The processing instructions should include
(a) A statement of the processing location and the principal equipment to be used
(b) The methods, or reference to the methods, to be used for preparing the critical equipment (e.g., cleaning, assembling, calibrating, sterilizing)
(c) Checks that the equipment and work station are clear of previous products, documents, or materials not required for the planned process, and that equipment is clean and suitable for use
(d) Detailed stepwise processing instructions (e.g., checks on materials, pretreatments, sequence for adding materials, critical process parameters (time, temperature, etc.)
The instructions for any in-process controls with their limits
Where necessary, the requirements for bulk storage of the products; including the container, labeling, and special storage conditions where applicable
Any special precautions to be observed

(c) The pack size expressed in terms of the number, weight, or volume of the product in the final container
(d) A complete list of all the packaging materials required, including quantities, sizes, and types, with the code or reference number relating to the specifications of each packaging material
(e) Where appropriate, an example or reproduction of the relevant printed packaging materials, and specimens indicating where to apply batch number references, and shelf life of the product
(f) Checks that the equipment and work station are clear of previous products, documents, or materials not required for the planned packaging operations (line clearance), and that equipment is clean and suitable for use
(g) Special precautions to be observed, including a careful examination of the area and equipment in order to ascertain the line clearance before operations begin
(h) A description of the packaging operation, including any significant subsidiary operations, and equipment to be used
(i) Details of in-process controls with instructions for sampling and acceptance limits

Pharmaceutical production processes classically are divided into steps or segments; each segment is to be checked and released during production. To provide a thorough proof that all production steps have been performed according to necessary quality requirements and have ended with the expected results, a batch-processing record has to be completed during production.

In Annex 6, the batch record is specified both for cylinders and gases to be delivered to a hospital (the second phrase leads to some misunderstanding, as it is not gas but cryogenic liquid that is supplied to a hospital tank:

Batch Record acc. GMP-Annex 6, pts 17/18.

	Data included in the record for cylinders/mobile cryogenic vessels		Records maintained for each batch of gas intended to be delivered into hospital tanks
a	Name of the product	a	Name of the product
b	Batch number	b	Batch number
c	Date and time of the filling operation	c	Identification reference for the tank (tanker) in which the batch is certified
d	Identification of the person(s) carrying out each significant step (e.g., line clearance, receipt, preparation before filling, filling, etc.)	d	Date and time of the filling operation
e	Batch(es) reference(s) for the gas(es) used for the filling operation as referred to in Section 22, including status	e	Identification of the person(s) carrying out the filling of the tank (tanker)
f	Equipment used (e.g., filling manifold)	f	Reference to the supplying tanker (tank), reference to the source gas as applicable
g	Quantity of cylinders/mobile cryogenic vessels before filling, including individual identification references and water capacity(ies)	g	Relevant details concerning the filling operation
h	Pre-filling operations performed (see Section 30)	h	Specification of the finished product and results of quality control tests (including reference to the calibration status of the test equipment)
i	Key parameters that are needed to ensure correct filling at standard conditions	i	Details of any problems or unusual events, and signed authorization for any deviation from filling instructions

j	Results of appropriate checks to ensure the cylinders/mobile cryogenic vessels have been filled	j	Certification statement by the Qualified Person, date, and signature
k	A sample of the batch label		
l	Specification of the finished product and results of quality control tests (including reference to the calibration status of the test equipment)		
m	Quantity of rejected cylinders/mobile cryogenic vessels, with individual identification references and reasons for rejections		
n	Details of any problems or unusual events, and signed authorization for any deviation from filling instructions		
o	Certification statement by the Qualified Person, date, and signature		

Quite a number of gas companies already use cylinder-tracking systems, for example, based on bar-codes read on the defined interface point of the production/manufacturing process. If this is followed, a major requirement would be fulfilled to document the individual identification of each filled cylinder (see Batch Release Record 95, p. 4; GMP-Annex 6 pts 17/18 and p. 126 in this book).

For integrated valves, some authorities want to have a record in the file – what kind of valve was used in a cylinder: simple valves can be identified by the manufacturer as well as the serial number of the valve, so this data would be sufficient.

5.3.5 Production

Production according to the GMP [87] follows the outlined frame of working in a QMS PQS, preferably according to EN ISO 13485 [80]. Starting with checked materials from known and assessed suppliers having all the necessary certifications documented, the materials are stored under defined and controlled conditions.

The manufacture is performed under strict control and documentation; labels are locked and secured, a traceability system in place. Before the manufacturing operation starts, it is checked that the equipment is within the specified limits of purity, functions, and that the operators are trained.

> *Before any processing operation is started, steps should be taken to ensure that the work area and equipment are clean and free from any starting materials, products, product residues, or documents not required for the current operation [87, p. 47].*

Cross-contamination is excluded by special measures as described in the Annex 6 and periodically checked. All quality relevant or critical steps of the production process are validated, new processes or new equipment have to undergo a management of change-process (MOC) and have to be released before going onstream by the pharmaceutical responsible person.

If intermediate products have to be prepared as it might be the case for some gas mixtures, it has to be taken care of thorough labeling and they have to be kept under appropriate conditions.

Packaging materials are to be properly stored; instructions exist to secure that the right labels are on the right place. The name and the batch number of the product shall be displayed at the production and/or packaging unit.

Finished products should be held under quarantine until the final release.

Rejected, recovered, or returned materials should be clearly marked and stored separately in restricted areas. Products returned from the market should be destroyed and not reprocessed, unless their quality is satisfactory without any cause for doubt.

At the time of the writing of this book, a revision of Chapter 5, Production, is in preparation. The new version puts special efforts on the danger of cross-contamination. In the future, the use of premises should be exclusively for medicinal products, handling of other products will be only allowed under "exceptional circumstances."

A second change affects the choice of the suppliers for starting materials. The level of supervision should be in relationship to the risks imposed by the individual material, taking account of the complexity of the manufacturing process. Especially mentioned is the manufacturer of starting materials, who should be fully integrated in the quality assessments of the receiver. The manufacturer of the finished product is responsible for ensuring that all described checks in the marketing authorization dossier have also been performed on the starting materials.

Another important change is the creation of the obligation of the market authorization holders (MAH) to ensure an appropriate and continuous supply of the medicinal product to their customers. In case the medicinal product ceases to be placed on the market, the MAH has to notify the competent authority in a sufficient time period before the cessation (except for unforeseen events). This point is highlighted in the discussion of the operation of Medical Gas Pipeline Systems (MGPS) in hospitals again.

5.3.5.1 **Annex 6**

All transfer of liquids and gases should avoid contamination. From the principle, liquids are more easily contaminated than gases, especially those gases liquefied under their own pressure, which might act as solvents under special conditions: super critical solvents like carbon dioxide; with some restrictions, nitrous oxide could also act as solvent. Pipes and hoses have to be purged with product gas before use; they should be equipped with nonreturn valves. They should also be equipped with product-specific connections to avoid cross-contamination.

As the liquid tank at the customer-site usually contains some remaining liquid cryogenic gas, refilling is allowed either with analyzed product or an analysis has to be made from the refilled tank after the procedure. Often the question is raised, if the liquid in the tank is homogenized during refill or not. In fact this is dependent upon several parameters, e.g. the flow of entering liquid in relationship to the size of the tank, the temperature difference between the residual liquid in the tank (comparably warm) to the entering liquid (comparably cold) and special measures to control the pressure in the tank before and during refill.

The set points can easily be distinguished:

- the flat bottom tank at the ASU, with a vast volume, where the flow of fresh and cold product slowly is entering, and
- the few cubic meter tank at the hospital quickly filled with a high velocity pump.

While in the first case the new product is deposited in horizontal layers in a calm environment, in the second case the pump is injecting a flow of liquid into the tank creating turbulence, intermixing the constituents of the tank very quickly.

Before filling cylinders or mobile cryogenic vessels, the batch (size, time period, etc.) should be determined and approved for filling.

In the context of Annex 6, the cylinder always is meant as the cylinder shell and the appropriate valve. If the valve is a simple one (just with or without an NR/RP cartridge) the batch record should note the manufacturer, the type of valve, and the serial number. If it is an integrated valve, to establish traceability, the individual valve should be recorded as they often have to go through a more detailed checklist during refill of the cylinder.

As the periodic test is carried out with water, this is a main source for contamination. For the periodic testing of medicinal cylinders, drinking water is to be used. Before the valve is (re)fitted, an internal inspection of the cylinder has to be made. Maintenance and repair of cylinders should only be done by the manufacturer of the medicinal product or by approved subcontractors of the manufacturer of the medicinal product (see GMP, Annex 6: it is never allowed to fill gas cylinder from other companies, if the conditions are not fixed in detail in a written contract.

Incoming cylinders should be checked for the basic properties: dents, arcs, burns, debris, and contamination with grease; and all other checks according to TPED (Transportable Pressure Equipment Directive) should be carried out. Cleaning should be done if necessary to comply with parts 9 and 45 of the Annex [85].

5.3.6 Quality Control

A new revision of Chapter 6 has been published and came into force by 1 October 2014 [88].

The independent action of quality control (QC) is an essential requirement to a satisfactory operation. QC is not restricted to the lab work but also an element of all decisions concerning the quality of the medicinal product. The primary task of QC is to set the points and values for sampling, testing the specifications, and recording all the gained data for further processing and evaluation. QC has to ensure that only satisfactorily tested material is used and that material that has not been released is not used for production.

The two key points are (i) QC is independent of other departments, especially the production department and (ii) the availability of adequate resources (in the sense of both adequate skills and training and adequate number). For satisfying both requirements, it is necessary to be able to perform checks and analyses with the necessary precision and reliability with respect to the market authorization.

The role of the Head of Quality (HoQ) had been described in the beginning of this chapter, see p. 109. A control laboratory should meet the requirements of Chapter 3 of the GMP and should be appropriately adopted to do the task to be performed.

External analysis capacity can be used, if the requirements of Chapter 7 "Contract Analysis" are met. The role of securing and supporting the efforts to reach pharmaceutical quality of the manufactured product is mirrored in the requirements for documentation and sampling as the key activities of the QC department (Table 5.2).

Table 5.2 Key features of QC [88, p. 2].

	The following details should be readily available to the Quality Control Department		The sample taking should be done and recorded in accordance with approved written procedures that describe
i	Specifications	I	i. The method of sampling
ii	Procedures describing sampling, testing, records (including test worksheets and/or laboratory notebooks), recording, and verifying	Ii	ii. The equipment to be used
iii	Procedures for and records of the calibration/qualification of instruments and maintenance of equipment	Iii	iii. The amount of the sample to be taken
iv	A procedure for the investigation of Out-of-Specification and Out-Of-Trend results	Iv	iv. Instructions for any required subdivision of the sample
v	Testing reports and/or certificates of analysis	V	v. The type and condition of the sample container to be used
vi	Data from environmental (air, water, and other utilities) monitoring, where required	Vi	vi. The identification of containers sampled
vii	Validation records of test methods, where applicable	Vii	vii. Any special precautions to be observed, especially with regard to the sampling of sterile or noxious materials
		Viii	viii. The storage conditions
		Ix	ix. Instructions for the cleaning and storage of sampling equipment

Table 5.3 Example for a protocol of a long-term stability program [88, p. 5].

Protocol for long going stability program	
i	Number of batch(es) per strength and different batch sizes, if applicable
ii	Relevant physical, chemical, microbiological, and biological test methods
iii	Acceptance criteria
iv	Reference to test methods
v	Description of the container closure system(s)
vi	Testing intervals (time points)
vii	Description of the conditions of storage (standardized ICH/VICH conditions for long-term testing, consistent with the product labeling, should be used)
viii	Other applicable parameters specific to the medicinal product

QC covers all used test methods, *including* those methods applied in the production. The entire test should be performed according to the methods approved by QC. Reference standards should be established as suitable. It is an important role of QC to establish a stability program, to find long-term changes in the quality of the product. The stability program should be defined and written according to Chapter 4 of the GMP, with publication of the results as a formalized report. The target is a regular check of important parameters (see Table 5.3) during the shelf life of the product, to ensure stability.

In case of more than one place of manufacture of the same medicinal product, it is important to take care of the validity and comparability of the different places. QC should check in defined periods of time the conformity of the used analytical equipment and the used methods with the market authorization. If a new laboratory or analytical equipment is to be installed, a technical transfer process has to be started.

The objective of such a process is to ensure that all equipment, operation conditions, used calibration gases, and the training of the personnel are according to the market authorization.

5.3.6.1 Annex 6

When extending this concept to the gases in cylinders, things become clearer (Table 5.4).

Reference and retention samples are not required for gases. If the initial stability studies have been replaced by bibliographic date (Ref), no ongoing stability studies are required.

5.3.7 Outsourced Activities

The holder of the marketing authorization is the responsible party in front of the authority [89]. Thus every contract, agreement, and arrangement has to respect the initial market authorization of the holder. Their PQS should include every outsourced activity and they should perform necessary controls and checks and

Table 5.4 Example of sampling plans for cylinder filling.

	Gas	Manifold	Sample plan	Determined
1	Single	Multi-cylinder	One cylinder for each manifold filling	Assay and identity
2	Single	One at a time	One cylinder per filling cycle	
3	Two or more gases	Manifold	Every cylinder of each component gas	Assay and identity
4	Excipients	Manifold	One cylinder per manifold filling cycle	Identity
5	Premixed gases		In-line testing, as under 1	Assay and identity
6	Premixed gases		No in-line testing, follow 4	Assay and identity
7	Stationary cryo vessel	Single	Sampling can be skipped, if delivery is accompanied by a certificate of analysis	
8	Mobile cryo vessel	Single	Every vessel	Assay and identity

provide help if needed. The contract acceptor must not subcontract any of their activities described in the contract to a third party, they should understand the content of the contract, they are not allowed to make unauthorized changes, and are subject to inspections by the competent authority.

The contract should describe clearly the outsourced activity, who is doing what, managing the technology transfer, supply chain, quality, and purchasing of materials and quality controls.

All the records related to the outsourced activities should be available to the contract giver.

5.3.8
Complaints and Product Recall

A system has to be in place to record all complaints and product information; these have to be reviewed carefully according to written procedures [90]. A major requirement is to avoid similar incidents from recurring. The requirement of introduction of the risk management principles has newly been entered in the last draft for revision. After an incident, the frequency and the severity of such an incident should be reviewed, thus giving the possibility of evaluating the necessity for risk mitigation measures.

When collecting all incidents and complaints, it is necessary to distinguish between quality issues and adverse medicinal effects. If the reporting does not show up a quality defect, the incident has to be recorded along the standardized procedure to collect all necessary data for further investigation and clarification. While defining the points to be collected for the standard procedure, the later investigation of an adverse event should therefore be facilitated. In the safety chapter of this book the pharmaceutical safety paragraph 7.1.4 also deals with Pharmacovigilance cases.

For the investigation of quality defects, at least the following points should be addressed (Table 5.5):

Detection or announcement of a quality defect should include the localization of the fault to decide between solitary defects, batchwise defects, or manufacturing defects affecting more than one batch. The investigation to distinguish between the different sizes of faults must occur in very short time, to fulfill the requirements of the authority for notification and to prevent harm from the patients.

When the root cause analysis has clarified the reason of the quality defect, it is now important to make decisions for the follow-up reflecting the level of risk (composed by severity and probability) for that quality defect and the consequences on the market authorization.

It is a matter of experience to know that information is not comprehensive in the early stage of investigation. In spite of that, risk-mitigating measures can be started: putting cylinders of the same batch into quarantine, stopping the supply of a batch selectively, collecting supplied but unused cylinders from the customers, emit a warning to the authority/health-care professionals, and make a recall.

Possible different markets have to be treated differently: a recall in a hospital is different from a recall at stockholders for medicinal gases, and this would be different from a recall at a health-care organization. In each case different measures have to be taken to avoid possible harm for the patient. If the recall takes more time than 24 h, the progress of the recall is to be documented in relation to the total number of cylinders in the same batch(es).

Table 5.5 Procedures for quality defect investigation [90].

i	The description of the reported quality defect
ii	The determination of the extent of the quality defect. The checking or testing of reference and/or retention samples should be considered as part of this, and in certain cases, a review of the batch-production record should be performed
iii	The need to request a sample of the defective product from the complainant and, where a sample is provided, the need for an appropriate evaluation to be carried out. The distribution information for the batch(es) in question. The assessment of the risk(s) posed by the quality defect
iv	The decision making process that is to be used concerning the potential need for risk-reducing actions to be taken in the distribution network, such as batch or product recalls, or other actions
v	The assessment of the impact that any recall action may have on the availability of the medicinal product to patients/animals in any affected market and the need to notify any such impacts to the relevant authorities
vi	The internal and external communications that should be made in relation to a quality defect and its investigation
vii	The identification of the potential root cause(s) of the quality defect
viii	The need for appropriate Corrective and Preventative Actions (CAPAs) to be identified and implemented for the issue, and for the assessment of the effectiveness of those CAPAs

Recalled products have to be kept under quarantine until the root cause is cleared and a formal decision involving the Qualified Person is made.

After investigation and clarification of the root cause of a quality defect, appropriate CAPAs have to be taken to defeat similar quality defects. It is part of the PQS to monitor and assess these measures for their effectiveness.

Any retrieval of the product from the distribution network should be necessarily classified as a recall. Management and notification of the authority brings self-confidence to the organization, as it trains the communication flows and reveals weaknesses.

In case of clinical trials, the information of all participants has to be executed as fast as possible. Here the time-consuming step is the unblinding of the products. The sponsor has to implement a process to rapidly identify the placebos, to be certain about the observed effects.

5.3.9
Self-Inspection

One of the most effective tools in the PQS is the Self-Inspection [91]. All quality-relevant matters of the organization, of the production, or of the quality assurance can be subject to Self-Inspection. A Self-Inspection plan should be set up to reach all parts of quality-relevant activities and to check their degree of conformity. Self-Inspections should be recorded to check the development and contain, if possible, agreed proposals for corrective measures.

5.4
Basic Requirements for Active Substances Used as Starting Materials (Part II of the GMP-Guide)

The development of the different directives concerning drugs and medicinal products showed up deficiencies in the quality treatment of the manufacturing of active ingredients [92]. Up to the 1990s the producers of active ingredients were not legally obliged to use any of the guidelines for good manufacturing so far developed.

In November 2000, the former Annex 18 to the GMP guide incorporated ICH Q7A for the manufacture of active ingredients used as starting materials and was used first on a voluntary basis by inspectors and manufacturers. With the amendments of 2004 (2004/27/EC and 2004/28/EC), the manufacturers were obliged to follow the GMP for the manufacture of active substances and to implement an appropriate QM system.

Table 5.6 is the guide as incorporated in Part II to elaborate the size of requirements for different types of starting points to produce active ingredients or APIs Typically, the manufacturing of the medicinal gases in an ASU is to be placed under the first line: chemical manufacturing.

In the revised version of 2010 the principles of risk management have been introduced.

Table 5.6 Application of Part II to API manufacturing [92].

Type of manufacturing	Application of this guide to steps (shown in gray) used in this type of manufacturing				
Chemical manufacturing	Production of the API starting material	Introduction of the API starting material into processes	Production of intermediate(s)	Isolation and purification	Physical processing and packaging
API derived from animal sources	Collection of organ, fluid, or tissue	Cutting, mixing, and/or initial processing	Introduction of the API starting material into process	Isolation and purification	Physical processing and packaging
API extracted from plant sources	Collection of plants	Cutting and initial extraction(s)	Introduction of the API starting material into process	Isolation and purification	Physical processing and packaging
Herbal extracts used as API	Collection of plants	Cutting and initial extraction	—	Further extraction	Physical processing and packaging
API consisting of comminuted or powdered herbs	Collection of plants and/or cultivation and harvesting	Cutting/ comminuting	—	—	Physical processing and packaging
Biotechnology: fermentation/cell culture	Establishment of master cell bank and working cell bank	Maintenance of working cell bank	Cell culture and/or fermentation	Isolation and purification	Physical processing and packaging
"Classical" fermentation to produce an API	Establishment of cell bank	Maintenance of the cell workbank	Introduction of the cells into fermentation	Isolation and purification	Physical processing and packaging

Increasing GMP requirements →

5.4.1
Annex 6

In addition to the outlined basic requirements of Part II, the active substance gases can be synthesized either by chemical synthesis or they can be obtained by exploitation of natural sources. The requirements as described in Volume 4, Chapter 7 of Part II (Materials Management) [93] do not apply for active substance gases:

- The manufacturer has to ensure that the quality of the ambient air is suitable for the air-separation process and does not affect the quality of the final active substance gas.
- The ongoing stability studies [94] have already been mentioned; they do not apply in case the initial stability studies have been replaced by bibliographic data, according to Note for Guidance CPMP/QWP/1719/00 [95]. Although this note is generated to specify the elements relating to the quality of medicinal gases in order to compile the documentation for Module III of the dossier, it gives completion of the more general rulings in Annex 6. As most of the statements refer to application for a marketing authorization, we have not taken up a detailed discussion of the document in this book.
- Requirements regarding reserve/retention samples [96] do not apply to active substance gases.

When active substance gases are produced through a continuous process (e.g., in an ASU), they should be continuously monitored for quality. Trends have to be recognized and evaluated.

In addition, transfers and deliveries of active substances in bulk should comply with those below for medicinal gases, see Sections 19–21 of Annex 6. Filling of active substance gases into cylinders or into mobile cryogenic vessels should comply with the requirements for gases as mentioned below (Sections 22–37 of Annex 6) as well as Part II, Chapter 9 (Packaging and Identification of APIs and Intermediates).

5.5
GMP-Related Documents (Part III of the GMP Guide)

Part III of the GMP guide contains a brief description of the necessary paperwork to support GMP activities and to make them visible for the inspecting authority.

5.5.1
Site Master File (SMF)

The SMF is prepared by the pharmaceutical manufacturer in order to give specific information about their GMP activities at a specific site to the authority [97]. It is one of the elements of general supervision at the authority and for the planning and undertaking of side inspections. To provide a quick overview, it should not exceed 25–30 pages written on A4 format in a readable size.

For the pharmaceutical manufacturer, the SMF is part of their PQS, carries dates for coming into force and for the next review and is kept updated accordingly. The general headlines are summarized in the Table 5.7.

Comments on explanations are to be found in the Explanatory Note [97].

Table 5.7 Contents of the Site Master File (SMF, acc. [97]).

1. General information on the manufacturer	
	1.1 Contact information on the manufacturer
	1.2 Authorized pharmaceutical manufacturing activities of the site
	1.3 Any other manufacturing activities carried out on the site
2. Quality management system of the manufacturer	
	2.1 The quality management system of the manufacturer
	2.2. Release procedure of finished products
	2.3 Management of suppliers and contractors
	2.4 Quality risk management (QRM)
	2.5 Product quality reviews
3. Personnel	
4. Premises and equipment	
	4.1 Premises
	4.2 Equipment
	– Cleaning and sanitizing
	– GMP-critical computerized systems
5. Documentation	
6. Production	
	6.1. Type of products
	6.2 Process validation
	6.3 Material management and warehousing
7. Quality control (QC)	
8. Distribution, complaints, product defects, and recalls	
	8.1 Distribution (to the part under the responsibility of the manufacturer)
	8.2 Complaints, product defects, and recalls
9. Self-inspections	
	Short description of the self-inspection system with focus on criteria used for selection of the areas to be covered during planned inspections, practical arrangements, and follow-up activities
Appendices	
Appendix 1	Copy of valid manufacturing authorization
Appendix 2	List of dosage forms manufactured including the INN-names or common name (as available) of active pharmaceutical ingredients (APIs) used
Appendix 3	Copy of valid GMP certificate
Appendix 4	List of contract manufacturers and laboratories including the addresses and contact information, and flow-charts of the supply-chains for these outsourced activities
Appendix 5	Organizational charts
Appendix 6	Layouts of production areas including material and personnel flows, general flow charts of manufacturing processes of each product type (dosage form)
Appendix 7	Schematic drawings of water systems
Appendix 8	List of major production and laboratory equipment

5.5.2 ICH-Q9 Quality Risk Management

The QRM explanatory note is the ICH Guideline Q9 on QRM [81]. QRM is part of the general approach of ICH to set up Guidelines ICH 8 (Pharmaceutical Development) [98], ICH 9 (QRM), and ICH 10 (Quality Systems) [82] on a new basis, the risk management view. For the pharmaceutical manufacturer, the implementation of a risk management–controlled quality assurance system offers the chance to introduce the experience of several years of manufacturing medicinal products, to reject search for highly improbable contaminations, and so on.

As a whole, the new concept offers improvement in the design of the quality process as it is

- based on scientific knowledge;
- enables continuous improvement;
- has greater transparency and efficiency;
- focuses on factors that add value for patients.

In summary, there is an improved relationship between industry and competent authorities on the basis of experience of the manufacturer and trust of the authority.

The ICH guideline Q9 references the known methods of error analysis and root cause analysis, so the manufacturer can use the whole contribution as a workbook to choose the appropriate method of fault analysis.

5.5.3 Q10 Note for Guidance: PQS

In the Note, the Guideline ICH 10 is referenced (see under PQS, Section 5.3.1) [82].

5.5.4 MRA Batch Certificate

Mutual Recognition Agreements (MRAs) describe processes to achieve uniform drug market authorizations (MAs). Once an MA is valid for more than country (especially third countries outside the European Union: Australia, Canada, Israel, Japan, New Zealand, and Switzerland), problems could arise that different forms with different statements about GMP conformity do not help reaching a situation of mutual understanding. By creating a standardized batch certificate, issued by the manufacturer in the exporting country, the flow of medicinal products could be standardized and regulated along the international agreements ACAA (Agreement on Conformity Assessment and Acceptance of Industrial Products) (Table 5.8).

Table 5.8 Contents of the Batch Certificate.[a)]

Contents of the batch certificate for medicinal products
1. Name of product
2. Importing country
[LETTER HEAD OF EXPORTING MANUFACTURER]
3. Marketing authorization number or clinical trial authorization number
4. Strength/potency
5. Dosage form
6. Package size and type
7. Batch number
8. Date of manufacture
9. Expiry date
10. Name, address, and authorization number of all manufacturing sites and quality control sites
11. Certificates of GMP Compliance of all sites listed under 10 or, if available, EudraGMP reference numbers
12. Results of analysis
13. Comments
14. Certification statement
15. Name and position/title of person authorizing the batch release
16. Signature of person authorizing the batch release
17. Date of signature

a) 1 June 2011 EMA/INS/MRA/387218/2011 Rev 5 Compliance and Inspection Internationally harmonized requirements for batch certification. In the framework of Mutual Recognition Agreements (MRAs), the Sectoral Annex on Good Manufacturing Practices (GMPs) requires a batch certification scheme for medicinal products covered by the pharmaceutical annex. Batch certification is also required in the Agreements on Conformity Assessment and Acceptance of Industrial Products (ACAA) and other appropriate arrangements on GMP between third countries and the European Union (EU).

5.5.5 Written Confirmation

Following Directive 2001/83/EC, active substances may only be imported from third countries into the European Union, if they are accompanied by a written confirmation of the competent authority of the third country, confirming the standards of GMP and that the controls in the manufacturing plant are equivalent to those used in the European Union.

6
Requirements of the New Good Distribution Practice (GDP)

6.1
Gas in Packages – No Difference from Other Medicinal Products

6.1.1
GDP – Targets and Tools

In 2013, the Commission released two versions of the guideline of Good Distribution Practice (GDP) of medicinal products for human use. The first one was released in March 2013 [99], revising the version of 1992 [100], and the second one in November 2013 [101], which completely replaced the first version. The target of the new regulation is the preservation of pharmaceutical quality throughout the complete supply chain, from the manufacturer to the pharmacist or to the hospital.

From the pharmaceutical point of view, after thorough development of drugs, validated manufacturing of both the active pharmaceutical ingredient (API) and the ready-for-use drug, distribution is the last part of the puzzle that is needed to complete an unbroken chain of traceability and quality checks.

Outside the purview of gases, it is not unusual to keep the transport of drugs under controlled conditions, for example, for insulin, for blood products, or vaccines. Here, the products have to be kept at a predefined temperature interval to prevent decomposition and to preserve the full efficacy of the drug when administered to a patient.

In addition, the number of falsifications on the market has risen substantially in recent years. Some of the newer European regulations [102] and directives deal with tamper-proof packages to prevent fraud during transport and repackaging.

Repackaging during transport by the intermediate partner is also nothing unusual in the real world. Often drugs are repackaged into smaller packages, which is a kind of manufacturing along the definitions of the directive.

All these processes have been widely activated with the local authorities, but in many countries, medicinal gases were exempted from these rulings. As a result, general rulings were frequently not respected from small transfillers, often in the public area, such as fire brigades, hospitals, medical device suppliers, among others.

Medical Gases: Production, Applications and Safety, First Edition. Hartwig Müller.

Furthermore, the gas industry itself has completely changed its logistics in the last 30 years. While the haulers of cylinders had stable and long-lasting contracts in the early 1970s, they were squeezed in their pricing in the meantime. To operate economically, they were driven to employ subcontractors for parts of the supply chain, for example, for small numbers of small cylinders (as characteristic for medicinal gases). The smaller the subcontractors, the more difficult it is to keep them acquainted to safety issues and to force them to apply safety rules, which is valid both for transport as well as pharmaceutical issues.

Since the 1980, when most of the bulk transports were performed by gas company–employed drivers and gas company–owned vehicles, the situation has changed completely: the whole process has been outsourced to "logistics-professionals," the drivers are employed by the hauler, the tractors are owned by the hauler, only the cryogenic tanks including their piping and pump for operations stay with the gas company. Here the common challenge is to get the drivers acquainted with the specialties of the cryogenic media (or cylinders) *and* to pharmaceutical questions, such as traceability and hygiene during loading and unloading besides the general issue of transporting dangerous goods.

In view of the gas logistics, it is now mandatory that the logistics company have to apply for permission at the authority and to name a responsible person for the transport of drugs. After the introduction above, it is clear that this new legislation will not be without consequences as the numerous subcontractors in the logistics as well as the drivers now need thorough training by competent management of the carrier company.

If this turns out to be difficult or impossible, alternative ways have to be found that are compatible with the new GDP which also includes the business of medical and medicinal gases. In the new regulation, the permission for wholesaling requires a quality management system (QMS) quite similar to that of the manufacturer of active starting materials. This includes a responsible staff and full documentation for the logistics of drugs. Part of the quality system is quality risk management, which allows to introduce scientific data about the risk according to guideline Q9 of the International Conference on Harmonization (ICH) [81] and to adapt the quality issues to the needs of the process.

In the following, the contents of the guideline will be explained briefly in the sequence in which they are mentioned in the guideline. The expression "wholesale distribution" is defined in the Directive [104] Art. 1,17 to be *"all activities consisting of procuring, holding, supplying or exporting medicinal products, apart from supplying medicinal products to the public".* ....

6.1.1.1 GDP – Quality Management

Wholesalers must maintain a fully documented QMS and provide adequate sourcing with staff and competent personnel and sufficient premises and equipment [105]. The QMS must comply with the processes of the wholesaler and provide the tools to manage processes such as outsourced activities, management review, and monitoring, so that the correct functions of the system can easily be checked Figures 6.1 to 6.3 showing the typical trucks for transport of a greater number of cylinders (6.1), for delivery of cryogenic liquid (6.2), and for single cylinders in the final customer delivery by a specialized contractor (6.3).

The implementation of an organization and the definition of appropriate corrective and preventive actions (CAPAs) to correct deviations and complaints and to prevent them in the future is a prerequisite for the permission of wholesaling medicinal products.

The basis for the implemented checks is the risk-based quality management. Risks should be identified via standardized methods and mitigated if the risk becomes too relevant either by probability or by severity.

6.1.1.2 **GDP – Personnel**

The wholesaler's responsibilities are supported by the personnel, working faithfully according to the operation procedures. Personnel and staff must be sufficiently competent to be able to carry out all delegated tasks [106].

A responsible person, appropriate organization and training, must be designated to support the personnel, so that the essential activities such as maintenance and development of the quality system, implementation, and maintaining of training programs, collection, and treatment of customer complaints, approval of suppliers, securing of contracted activities to be in-line with GDP, performance of self-inspections, and other important processes run smoothly.

Hygiene should be appropriately organized.

6.1.1.3 **GDP – Premises and Equipment**

The premises must be suited for the storage and handling of drugs and proper storage of the products must be ensured. They should be clean, dry, and maintained within acceptable temperature limits. Cleaning programs should be in place [107].

Unauthorized persons must not have access to the premises.

Storage and storage areas should follow the flow of incoming and outgoing products and be clearly separated to avoid any mismatch between ordinary products, returned products, complaints, and other quarantined goods.

Premises should be free from insects, rodents, and other pests.

Equipment should be designed to suit the intended purpose. Equipment used to control or monitor environment for storage should be calibrated at regular time intervals.

6.1.1.4 **GDP – Documentation**

Documentation is the core of most of the good manufacturing practice (GMP)-compliant activities. It should be written in a language understood by the personnel. Procedures should be signed and dated by the responsible person and other documentation should be signed and dated by the authorized person. The title, nature, and purpose should be clearly stated and they should be kept up-to-date; a version control system should be applied to procedures.

Employee should access those papers necessary for them to do their work appropriately.

Figure 6.1 Cylinder truck.

Figure 6.2 Bulk tanker (own Picture: Air Liquide Healthcare).

Figure 6.3 Subcontractor's truck (Picture: Santrans GmbH, Germany).

Records should be made at the time of an operation and should cover the relevant activities, for example, delivery notes and purchase/sales invoices for any transaction of medicinal products received, supplied, or brokered.

6.1.1.5 **GDP – Operations**

The wholesale distribution of medicinal products in the European Union comprises all products with a market authorization by the Union or by a member state. Wholesalers must obtain the medicinal products only from persons in possession of a wholesale permission or a manufacturing authorization, covering the product in question.

Another important point is the obligation for the wholesaler to carry out a "due diligence" when entering in a contract with a new supplier, in order to assess the suitability, competence, and reliability of the other party.

At the same time, the wholesaler must ensure delivery of medicinal products only to persons in possession of a wholesale license or to those who are authorized to deliver medicinal products to the public.

6.1.1.6 **GDP – Complaints, Returns, Suspected Falsified Medicinal Products, and Medicinal Product Recalls**

Handling of the medicinal products is the key role of the wholesaler. All activities must be systematically defined, recorded, and carefully handled.

There should be a thorough treatment and segregation of complaints and returns by an appointed person with sufficient support.

Identification and information of the authority and the market authorization holder about falsified products must be carried out according to written procedures.

There should be annual evaluation of the effectiveness of the product recall organization.

6.1.1.7 GDP – Outsourced Activities

Wholesale activities cover the entire logistic process chain. Especially the parts near to the customer are those relevant for outsourcing. The wholesaler can outsource these processes only after exact definition (what is to be done), framing of conditions and legal obligations (how to preserve the integrity of the product), the necessary documentation (how to sustain traceability), and last but not least, the necessary controls and checks of the activities exercised by the contractor (Figure 6.3).

6.1.1.8 GDP – Self-Inspections

Self-inspections have proved to be an efficient tool to identify deviations in compliance in the daily routine, if conducted by trained and competent personnel. They cannot be replaced by external audits.

The subjects for self-inspections should cover the activities under the regulations, the GDP, and the implementation. The different subjects can be followed up in a self-inspection program within a defined time frame.

Records of self-inspections should be retained.

6.1.1.9 GDP – Transportation

When planning transportation, a risk-based approach should be used. It is the objective and responsibility of the supplying wholesaler to preserve the integrity of the product (breakage, aging, theft) and to demonstrate that the conditions for the transport always have been maintained.

The vehicles used must be operated and maintained along written procedures, including cleaning and safety precautions.

The containers must be equipped with appropriate labeling, carrying the necessary information for handling and storage of the products. The contents and source of the contents should be identifiable by the labels used.

6.1.1.10 GDP – Specific Provisions for Brokers

The GDP Guideline follows the definition of a broker in [108]: "a 'broker' is a person involved in activities in relation to the sale or purchase of medicinal products, except for wholesale distribution, that do not include physical handling and that consist of negotiating independently and on behalf of another legal or natural person."

6.1.1.11 GDP – Conclusions

The processes in the pharmaceutical industry have reached a kind of maturity in the last years. Under the rising cost pressure of the development of new medicinal products, the pressure of the authorities to prove efficacy and an explicit benefit of the new drugs in comparison to the existing ones developed structures to avoid excessive costs.

This was one of the reasons that logistic processes have gained relevance: manufacturing sites employing large numbers of people have been cut down to highly specialized units for worldwide manufacturing of a small spectrum of similar drugs.

The logistics itself had to support these efforts for minimal costs, with the result that besides the manufacturing of the API and the final medicinal product, new risks came up: long transportation distances are not usually managed by one hauler; often a number of haulers with subsequent activities compose the supply chain for a medicinal product. Erosion of initial quality can be observed: the smaller the subcontractor, the higher the risk that the common safety standards can be violated.

The new GDP shows the weight that is laid by the authorities on this part of the pharmaceutical supply chain. A bouquet of measures is created in order to fence the logistics activities and not to lose the quality that has been gained through the regulated manufacturing process. To exhaustively comment all these activities laid down in the new regulations, the gas industry has to make the appropriate experience first.

7
Safe Handling of Gases

7.1
Safe Handling of Gases

Numerous papers have been written about safety, in general, and also about the safety of gases, in particular. In modern industrial approaches, risk management is a key technique to review inherent dangers: typical hazards of installations, techniques, or procedures are identified and assessed in view of their probability and severity.

This practice offers us a chance in this book to use the long years of experience of the gas companies to go for the actual threats and not so much for those of the past. According to the more practical targets of this book, we will try here a complementary approach, using the knowledge of the most common errors that occur in daily life. The following material should not replace the very detailed material contained in the various web-sites of ICH and in the numerous standards about that topic, it should rather be used as a simplified access to a vast reservoir of introductory resources.

7.1.1
Hazards and Risks

We begin with a few words to improve common understanding of hazards and risks: hazards generally describe possible perils during the use of dangerous goods or installations. A risk is described in two dimensions: the probability and the severity [109] and a similar approach can be found to estimate pharmaceutical quality risks in ICH Q9 [81].

Risk is a quantitative measurement of the impact of a potential event and its probability. To yield a reliable estimation, a risk management approach has to be performed. Thus a risk gives a quantitative picture of a hazard, including both the probability of the occurrence of an event and its possible severity (Tables 7.1 and 7.2 as of 118a).

The common approach includes already a standardized risk management approach as described in the European standards (e.g., EN ISO 14971 for Medical Products and the ICH Q9.). To estimate the real threat of a hazard it is necessary

Medical Gases: Production, Applications and Safety, First Edition. Hartwig Müller.

Table 7.1 Risk evaluation matrix – **probability** – keywords in bold [109].

Category of probability	Category name	Frequency (per year)	Explanation
0	**Improbable**	$P < 10^{-6}$	There are no known events of this kind
1	**Very seldom**	$10^{-6} < P < 10^{-5}$	Event requires a combination of rare events
2	**Seldom**	$10^{-5} < P < 10^{-4}$	Event has occurred once in a similar environment
3	**Possible**	$10^{-4} < P < 10^{-3}$	Event has occurred once during the lifetime of the equipment
4	**Often**	$P > 10^{-3}$	Event has occurred several times during the lifetime of the equipment

Table 7.2 Risk evaluation matrix – **severity** – key words in bold

Severity level	(Occupational) safety	Environmental safety	Installation safety
0	No injuries; **negligible**	No exterior damages	No damage
1	Minor injuries; **marginal**	Moderate damages w/o long-term effects	Light damage to installation
2	Serious injury; **serious**	Serious damages, can be corrected	Important damage of the installation and the building
3	Potential fatality; **critical**	Serious and long-term damage	Serious damage of the buildings, installation, and neighborhood
4	Major accident, several fatalities; **catastrophic**	Ecological disaster	Massive destruction of buildings and installations and of the neighborhood

to rate both the severity and the probability of a feared event. To avoid arbitrary or influenced rating, it is important to use existing reliable statistical data to receive a clear picture of the true probability of a feared event and the most probable severity.

The complete risk management process includes

- risk identification
- risk assessment
- risk mitigation
- evaluation of the residual risk
- experience feedback.

For all of these methods exist standardized approaches for analyses, as the Ichikawa analysis, hazard identification and assessments etc. In the ICH website a very instructive 'Briefing Pack' helps to get acquainted with the procedure. The

Table 7.3 Example of an assessment matrix [109].

		Severity				
		Negligible	Marginal	Serious	Critical	Catatstrophic
Probability	often					
	possible					
	seldom					
	very seldom					
	improbable					

outcome of a risk management process is a table, where all the results have been collected in the form of a diagram (see Table 7.3). The different grades of severity appear on the x-axis and the different grades of probability are outlined on the y-axis.

The red fields show risks that are not tolerable and which have to be defeated by measures. The yellow fields show risks which might be tolerable or where first mitigation measures alone do not succeed. Green fields show risks sufficiently mitigated and defeated by appropriate measures. By following the risk management process, the red fields have to be eliminated by risk mitigation measures. Usually, operation is possible only when the risk mitigation leaves and green fields occupied. Yellow fields, if present, require urgently mitigation measures before going into operation.

Following that approach, the accident statistics of the gas companies collected by the EIGA (European Industrial Gas Association) are the most valuable tool to realistically estimate the probability of occurrence of a possible event.

From that viewpoint, we have known for years that the most probable accident with a gas is due to the cylinder falling down and crushing toes and fingers, because of its high weight.

Shearing off of a valve of a falling cylinder seldom occurs owing to mounted valve-saving accessories such as caps or permanent valve guards. Only small cylinders in the hospital environment in many countries carry no additional valve guards and account for a considerable number of falling accidents, very seldom with the loss of the valve. In this case, the energy load of the cylinder pressure is released and the cylinder is propelled through the room. Shearing off of the valve has not been often reported as the medical cylinder (2 and 3 l volume) itself is not very heavy, so in most cases, the valve is only bent or the hand-wheel is broken when the cylinder falls to the ground without the release of gas.

Cryogenic liquids are often underestimated in view of their effect on biological materials, a major reason for that being the so-called Leidenfrost phenomena: the temperature gap between the cryogenic liquid and the wetted surface is so high that a vapor layer is formed around the droplets of the liquid, isolating for a short time the surface from the impact of the cryogenic temperature, thus creating an effect of "nothing happens". In fact, there something does happen: while tissues very often suffer from cold burns, all other common materials slip into the danger of cold embrittlement: only few alloys can withstand the cryogenic temperature without significant alteration in their crystal structure: losing resilience completely and breaking like glass.

In the following sections, we will go through the most common hazards when handling gases or cryogenic liquids. By rating the probability of a feared event, preventive measures can be developed; by observing the severity of a feared event, protective measures can be used to mitigate the consequences of such an event. Both help to generate a feeling of consciousness while handling compounds with specific properties. Further information and training of the methods can be found in the sources discussed in the following [111–113].

7.1.2 Safety and the Pressure Container

Pressure vessels are uncommon containments in the healthcare environment. Everybody working with pressure vessels (i.e. transporting, connecting, or applying the gas) should have attended appropriate training to do a safe job. Fundamental information about all properties of the gas and the containers is supplied by the manufacturer/supplier. A golden rule is never to try to fix any hardware problem on your own, always rely on the expertise and skill of the gas supplier or the manufacturer of the hardware installation. From the unwanted venting of gas from pressure receptacles a specific hazard is generated that can devastate the whole working environment, and additionally can establish life-threatening situations by propelling of heavy objects through the air.

The predominant attribute of a transportable pressure vessel is its weight. Starting with small (2–5 l) to medium size (8–12 l) and big size (40–60 l) cylinders, the user is confronted with considerable tare weights of the cylinders (Table 7.4).

Depending on the use of the cylinders, the weight also is a strong argument for a good organization of the gas containers in a clinical environment. Big cylinders (approximately 40–50 l) are exclusively used to feed central medical gas supply units, while small cylinders are mainly used in the vicinity of the patients, for example, in case of transport supply for one patient preferably a versatile 2-l cylinder is used, attached to the patient's bed.

7.1.3 Main Technical Risks

7.1.3.1 Storage

Storage and use of cylinders must follow national regulations such as (i) there is no storage below ground level, only in well-ventilated areas; (ii) limited number of cylinders to be used at the application or at the feed; (iii) cross-contamination of gases is avoided; and (iv) full and empty cylinders are separately stored.

All cylinders, stored and in use, including those in the central feed of MGPSs (medical gas supply systems), have to be secured with chains or equivalent measures must be taken to avoid their falling down. To avoid finger irritations one should carry appropriate gloves when handling cylinders and wear safety shoes to

Table 7.4 Tare weights and dimensions of commonly used cylinder sizes[a)]

Steel cylinders[b)]

Volume (water capacity, l)	Filling pressure (bar)	Gas volume (15 °C, 1 bar) (m^3)	Diameter (mm)	Length (mm)	Tare weight (kg)
1	200	0.2	83	275	1.45
2	200	0.4	100	350	2.5
5	200	5	140	450	7.2
10	200	2	140	820	12
10	300	3	140	845	19
33	300	10	229	1060	54
40	150	6	204	1615	73
50	200	10	229	1515	65
50	300	15	229	1540	73

Aluminum cylinders[c)]

Volume (water capacity, l)	Filling pressure (bar)	Gas volume (15 °C, 1 bar) (m^3)	Diameter (mm)	Length (mm)	Tare weight (kg)	Al-alloy
1	200	0.2	102	240	2	AA6061
2	200	0.4	117	336	3.2	AA6061
2	200	0.4	102	360	2	AA7060
10	200	2	140	970	12	AA6061
11	200	2.2	180	624	11.2	AA7060
40	200	8	229	1455	46	AA6061

a) Reference [14].
b) Reference [115].
c) Reference [116].
The stated values are average values. For a cylinder equipped with a valve, the weight of the valve (approximately 0.5 kg) has to be added to the weight of the cylinder to arrive at the total tare weight.

avoid being hit by falling cylinders during transport. The big 50-l cylinders especially are difficult to handle when moved out of the feeding rack.

7.1.3.2 **Transport**

As already mentioned, the main safety risk of a cylinder is falling down of the container. As a consequence, appropriate carts have to be used for transport of single cylinders; for more than one cylinder, overpackings [Technical expression from ADR] such as boxes or palettes are quite common.

If it is necessary to transport full gas vessels in small cabinets (e.g., cars or lifts) the gas content of the gas vessel should be kept in mind. If problems arise, for example, if the elevator stops and one of the cylinders might leak an asphyxiant gas, the danger is obvious. One should never use the same elevator as the cylinder at a time. In a car, all the windows should be kept open to allow for circulation of air.

Small cylinders should never be carried by the top of the valve or hand-wheel. The cylinder is comparably heavy and the torque is big enough to easily open the head-valve of the cylinder, and the consequence would be a sudden release of gas.

A more sophisticated hazard is the leak at a cylinder with liquefied gas: when the cylinder is stored in a way that liquid phase covers the valve outlet in the cylinder (e.g. in horizontal or top-down position), liquid will be vented to the outside. When released, the liquid instantly vaporizes and cools down the valve body considerably, so cold burns could happen when somebody intends to close the leaking valve.

7.1.3.3 Application

The bed of a patient should never be used for temporary storage or use of cylinders. One must also take care of attached pressure regulators, which could break, should the cylinder fall down.

If cylinders are used during transport of patients (e.g., from one building to another), they have to be attached firmly to the bed. These cylinders have to be kept connected, as the patient needs the oxygen during transport, so special care is advised.

7.1.4 Pharmaceutical Safety

Pharmaceutical safety has a twofold meaning: on the one hand the pure technical procedure along the GMP rulings and confirming all production steps, their checks and their recording and on the other hand the pharmacological effects on the big number of patients on the market, subsummized under the topic pharmacovigilance.

Pharmacovigilance is part of the general strategy of risk mitigation measures, but in opposite to the technical view pharmacovigilance is focused on the pharmacological effects of the drug administered to a patient. Here the target is the continuous detection, announcement and recording of adverse reactions or side effects associated to the contact with the drug and their impact on the patients once it has been placed on the market. This being a prerequisite to define, to plan and to implement risk mitigation measures and to assess their effectiveness related to the pharmacological efficacy of a drug [110].

Pharmacovigilance activities include: [110]

- Collecting and managing data on the safety of medicines
- Looking at the data to detect 'signals' (any new or changing safety issue)
- Evaluating the data and making decisions with regard to safety issues

- Pro-active risk management to minimise any potential risk associated with the use of the medicine
- Acting to protect public health (including regulatory action)
- Communicating with and informing stakeholders and the public
- Audit, both of the outcomes of action taken and of the key processes involved.

Those directly involved in pharmacovigilance include:

- Patients who are the users of medicines
- Doctors, pharmacists, nurses and all other health care professionals working with medicines
- Regulatory authorities, including the European Medicines Agency (EMA) and those authorities in the Member States responsible for monitoring the safety of medicines
- Pharmaceutical companies and companies importing or distributing medicines.

Since 2004 the commission has spent big efforts to improve the situation, resulting in a number of regulations: Directive 2001/83/EC (18) and Regulation EC 726/2004 marking the legal framework, while regulation EU 520/2012 describes the operational details for the marketing authorization holders and the national competent authorities as well as the EMA [103a,b]. Additionally the 'good pharmacovigilance practices guidelines (GVP)' [110] replace the preceding rulings in Eudralex Vol 9A-pharmacovigilance.

As this book is more oriented to the technical handling of gases and the related risks, treatment of the pharmacological approach would extend the scope in a presently unwanted extent, although the regulations presently being in the implementation phase in Europe [114].

In the more technical view pharmaceutical safety has to be secured on different levels. First of all, the integrity of supplied cylinders for pharmaceutical use must be checked: are all necessary labels applied and do the batch numbers on the cylinders correspond to the batch numbers in the delivery sheet? Are all labels readable and in good shape? Second, what is the all-over impression the cylinders present: no one would expect a cylinder to be clinically clean, but if the cylinder's paint is scratched and the surface rusty, it is recommended that they are not used in high-risk areas, such as an operation theater. There is no other alternative than to refuse cylinders that do not correspond in their description to their later clinical use which sometimes cannot be deducted from returning empty cylinders as Figure 7.1 is showing.

The other level of safety is a quality check itself. The supplier/manufacturer guarantee the correct quality with their name. In spite of that, in some European countries, national laws prescribe the test of a sufficient number of cylinders, to be sure that conditions at the manufacturer to produce drugs are followed as described in the relevant pharmaceutical dossier. It is recommended to follow up a test plan to be sure that over the year all necessary testing has been performed and documented.

Figure 7.1 Returning cylinders.

As in most cases, the pharmacist in the hospital does not have sophisticated analytical instrumentation at hand; it is therefore advantageous to use the methods of the Pharmacopoeia dedicated to the users – for the gases in most of the cases, these are simple methods with test tubes.

In many European countries, the spreading of unwanted germs in hospitals is an issue. Therefore, it might be a good idea to identify those areas in the hospitals, where there is a high probability that germs like MRSA (Methicillin-resistent Staphylococcus aureus) or other such highly infectious germs are present. Here all cylinders entering and leaving the premises should be subject to extended housekeeping checks: documentation of which cylinders go in, which cylinders are returned, and definition of a procedure to treat the surface of the cylinders before returning them to the supplier securing that the cleaning procedure has been carried out and finally recorded.

Comparing the measurement of germs on the outside of a cylinder and those of the gas show clearly that the contamination risk lies on the settlement of germs on the outside of the cylinder rather than in the inside. During the normal operations of manufacturing gases, pressure changes are quite common and these changes reduce germs very effectively; all measurements in the gas back that approach.

The most possible sources for contamination of the gas or of the patient are situated in the moistener flask or on the outside of the cylinder.

7.2 Safety and Pressure

Once the cylinder is brought to position, it is time for connection. All valves have to be kept oil- and grease-free to avoid flash fire when coming into contact with oxygen. Always open the (cylinder) valves *slowly* and by hand.

The commonly used cylinders contain up to 200 or 300 bar pressure, which is a considerable amount of stored energy: a mass of 1 ton falling down 200 m equalizes the energy stored in a 50 l cylinder, or, chemically, about 0.5 kg of TNT (trinitro-toluene, a common industrial explosive) contains the same energy as this cylinder.

This means that every cylinder represents a considerable amount of energy (about 0.5 kWh), which is comparable to that of a loaded spring. Handling of cylinders should also take into account of the fact that every unwanted and uncontrolled venting of a cylinder can lead to serious incidents, even when a cylinder is propelled through the room (the other possibility is that the connection pipe – pigtail – bends around like a whip, creating bad injuries).

Cylinders always have to be transported in safe condition: the valves should be closed and the valve guard attached; the cylinders should be secured against falling; withdrawal of gas should be done only with attached regulator; and the cylinders should never be interconnected to each other without a backflow resistor.

In all cases, an appropriate regulating device has to be used to withdraw gas from the cylinder; a cylinder should never be used without pressure regulator to tap gas. Cylinder valve threads are deliberately not equal and only matching device should be used. The cylinder thread is a safety function: all left-handed threads contain flammable gases, while right-handed threads are used for all other gases (refer to Table 2.1 in Chapter 2 of this book).

Although different gases have different threads, the user might get the impression that a nonmatching valve connection will be matching by bad tolerances. So, when first using a regulator care has to be taken to take the right connection as indicated in the list. Before attaching a regulator check the maximum allowed pressures, do they match with the pressure of the gas?

The reading of the pressure at the gauge of the regulator after opening of the cylinder is not always a valid indicator for the amount of gas in the cylinder; if the gas is a permanent gas: the critical point is low enough to prevent condensation in the cylinder (precipitation of liquid phase) under filling pressure at ambient temperature. If the gas is liquefied under its own pressure, we find a liquid phase in the cylinder, which feeds the gas phase to the equilibrium pressure at ambient temperature (see Chapter 2 of this book). The pressure above the liquid in the cylinder is only dependent on the temperature of the liquid; under moderate withdrawal, it remains constant until the last drop of liquid has vaporized, then the cylinder pressure breaks down immediately.

In case of large consumption of the gas, this would lead to a strong vaporization of the liquid, cooling the liquid down, until the vapor pressure collapses: the withdrawal then decreases with the falling temperature of the liquid and it finally stops.

7.3 The Compound's Chemical Properties

Until the late 1970s, cylinder gases were only distinguished by a cryptic system of colors (nearly each European country used its own system) and almost no further information other than a permanent stamping in the cylinder neck provided safe working with a gas.

Again, a European directive 67/548/EEC [117], and the subsequent 99/45/EC [118], cleared things to a satisfactory degree. Today, all gas cylinders come along with readable labels, standardized hazard and risk phrases, and standardized classification and thus with consistent dangerous goods/dangerous substances warning symbols. This is due to the GHS, the global harmonized system finding its precipitation in the European CLP (Classification, Labeling, and Packaging)-regulation 1272/2008 [119].

On the label of the gases, a first classification is indicated by the signal word "WARNING" or "DANGER" together with the precautionary (P-) statements and hazard (H-) statements, depending on the classification under the regulation.

In addition to these explanatory statements, the label indicates the main danger with the international GHS-symbol, so the user is easily informed about the inherent hazards accompanying a cylinder gas. An example for the new design of cylinder labels is given below (Figure 7.2).

Moreover pushed by REACH (REGULATION 1907/2006 concerning the Registration, Evaluation, Authorisation and Restriction of Chemicals (REACH)) [122], for every chemical substance marketed in the European Union, the supplier is obliged to complete a dossier about the hazards and properties of classified chemicals (depending on the dangerous goods classification and the amount of substance brought on the market). The supplier is obliged to supply a safety data sheet (SDS) with the first delivery and afterward, repeating this to the professional customer, upon important changes.

Figure 7.2 Cylinder label according to CLP [119] CLP-Symbols are replaced by ADR-placards.

Table 7.5 Important H- and P-statements for common gases in German, English, and French.

H220	Extrem entzündbares Gas	Extremely flammable gas	Gaz extrêmement inflammable
H221	Entzündbares Gas	Flammable gas	Gaz inflammable
H271	Kann Brand oder Explosion verursachen; starkes Oxidationsmittel	May cause fire or explosion; strong oxidizer	Peut provoquer un incendie ou une explosion; comburant puissant
H272	Kann Brand verstärken; Oxidationsmittel	May intensify fire; oxidizer	Peut aggraver un incendie; comburant
H280	Enthält Gas unter Druck; kann bei Erwärmung explodieren	Contains gas under pressure; may explode if heated	Contient un gaz sous pression; peut exploser sous l'effet de la chaleur
H281	Enthält tiefkaltes Gas; kann Kälteverbrennungen oder -verletzungen verursachen	Contains refrigerated gas; may cause cryogenic burns or injury	Contient un gaz réfrigéré; peut causer des brûlures ou blessures cryogéniques
P220	Von Kleidung/ … /brennbaren Materialien fernhalten/entfernt aufbewahren	Keep/store away from clothing/ … /combustible materials	Tenir/stocker à l'écart des vêtements/ … /matières combustibles
P244	Druckminderer frei von Fett und Öl halten	Keep reduction valves free from grease and oil	S'assurer de l'absence de graisse ou d'huile sur les soupapes de réduction
P370	Bei Brand:	In case of fire:	En cas d'incendie:
P376	Undichtigkeit beseitigen, wenn gefahrlos möglich	Stop leak if safe to do so	Obturer la fuite si cela peut se faire sans danger
P403	An einem gut belüfteten Ort aufbewahren	Store in a well-ventilated place	Stocker dans un endroit bien ventilé

On the basis of the SDS, the professional customer is able to make a risk assessment and write handling advice for safety of working with the specific gas to be hung at the working place, where the gas is handled. The decision regarding what kind of personal safety accessories (PSA) should be worn at a specific working place is also based on the SDS. Table 7.5 presents a compilation of the most common handling instructions for gases.

The European gas industry contributed to this by the generation of a unique color code system, consisting of different coloring of the head and the shoulder of the cylinders. Following this system (EN 1089-3), the color of the shoulder indicates the main hazard when working with that gas, for example, oxidizing, flammable, or toxic. The body of the cylinder can be painted in a neutral color, which can be individually chosen by the gas manufacturer, for example, giving organizational distinction to separate technical from special gases.

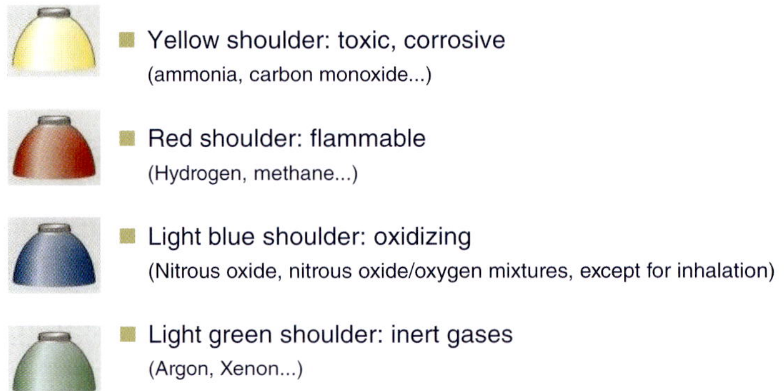

Figure 7.3 Cylinder colors according EN 1089-3 indicating the primary danger.[1]

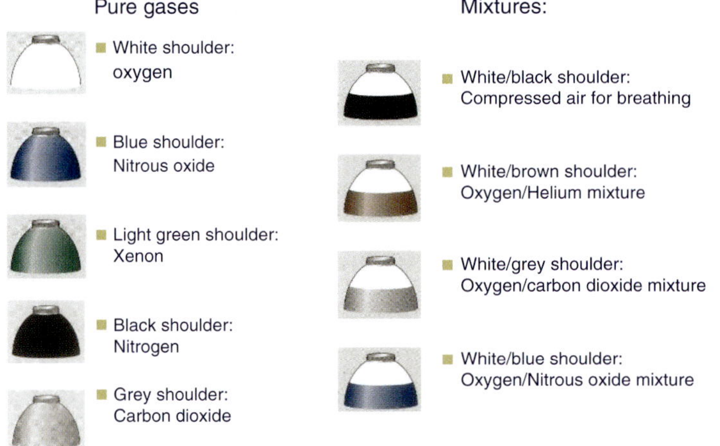

Figure 7.4 Cylinder colors according EN 1089-3 for medicinal gases (cylinder always white).[1]

For medicinal gases, the approach is a little different: to distinguish medicinal cylinders easily from others, the body of the cylinders is always painted white (Figures 7.3 and 7.4).

The color of the shoulder indicates either the gas, or, in case of mixtures, the balance gas. For the first time, medicinal gases now can easily be separated from technical gases; on the other hand, the white color is not the best choice for cylinders in

[1] Flyer of IGV, Color codes for Industrial and Medicinal Gases.

heavy duty, for example, in ambulance cars or first-aid bags. Due to the medicinal use of the cylinders some of the competent authorities in some of the European countries had decided to ask the manufacturer for additional information. This is different in the countries and might even repeat points from the patient's leaflet which is accompanying every cylinder.

7.4 Cryogenic Liquids: Low Temperature and Vast Development of Gas

The main application of cryogenic nitrogen in health-care environments is storage of biological samples and tissues. From our own experience the inherent and most obvious hazards are often underestimated owing to the special side effects of the cold liquid.

Liquid nitrogen (LIN) behaves like a liquid, but most of the gloves worn just isolate against contact with cold solids, liquids passing without resistance through and burning all the tissue it comes in contact with. This effect can easily be compared with a pot cloth, wetted by hot water and immediately losing its isolation ability.

The inherent low temperature of the cryogenic liquid typically is a cause for embrittlement. Conventional materials become brittle after contact and after only a short time, the material (most commonly the ground, as the example in Figure 7.5 shows) will erode and lose its stability. The effect is dramatic for

Figure 7.5 Example for erosion after contact with cryogenic temperatures.

Figure 7.6 Warning sign: asphyxiation.

pressurized installations, which come into contact with the cryogenic temperature incidentally: if not designed to resist, a drastic reduced pressure resistance is the result, with possible serious consequences [120].

Freezing of biological tissues with LIN is often supported by programmable device realizing a neatly defined temperature program to avoid the generation of ice crystals which would damage the cells' walls. In easier designed experimental environments, the freezing is done in small labs, often without sufficient ventilation.

Special care has to be taken in the vicinity of storage containers to prevent hazards of LIN, especially in confined spaces. The vaporizing liquid develops upon evaporation about 850-fold volume of gas (!), creating a serious danger of asphyxiation, as the oxygen content of the breathing air will be rapidly decreasing in the vicinity of the container/s (Figure 7.6).

This risk is extremely high when large amounts of substance are to be cooled down, thus warming up the liquid, generating large amounts of gaseous nitrogen, or during cooling down of "warm" containers, when all LIN in the containers has been removed by previous operations. Until the container has reached cryogenic temperatures again, LIN is vaporized in big quantities. As an example, Figure 7.7 shows oxygen concentrations in an open cryo cabin.

The person concerned has only minimal or no chance to escape: nitrogen is invisible, has no smell, no obvious warning signs appear, when working in oxygen-deficient areas:

Asphyxia – Effects and Symptoms of Reduced O_2 Concentration (vol%) (after EIGA [111])

18–21%:	No discernible symptoms detected by the individual.
11–18%:	Reduction of physical and intellectual performance without the sufferer being aware.
8–11%:	Possibility of fainting within a few minutes without prior warning. Risk of death below 11%.
6–8%:	Fainting occurs after a short time. Resuscitation possible if carried out immediately.
0–6%:	Fainting almost immediate – Brain damage, even if rescued.

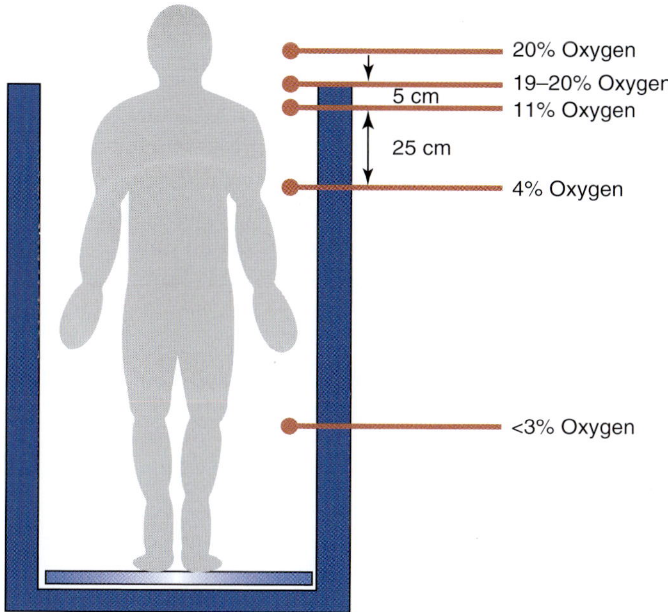

Figure 7.7 Nitrogen enrichment near the evaporation source (EIGA, [111]).

This is also the place to make a point about carbon dioxide: carbon dioxide is often seen as an asphyxiant only, so to be secure, only oxygen concentrations are measured at the exposed position. Carbon dioxide has a systemic effect on the breathing [121] of the individual. The concentration of carbon dioxide controls the gas exchange via different parameters, such as the frequency of breathing, the pulse rate, and the blood pressure. If the content of carbon dioxide in the breathing air is too high, the system is compromised, and the organism is intoxicated with carbon dioxide.

> Carbon dioxide is not just an asphyxiant!

After only 30 min of breathing, 4–5 vol% of carbon dioxide the respiratory center is heavily stimulated and signs of intoxication (headache and loss of judgment) will occur. It should be always remembered that this effect will be observed even if the oxygen concentration is above the critical limit of 19 vol%.

Cryogenic temperatures on their own will create another risk, the danger of cold burns after contact with liquid or with cold parts of the piping or manifold. The misleading interpretation of the Leidenfrost phenomena has already been mentioned in the introduction. In addition, when the skin comes into contact with a cryogenic liquid, the development of pain takes place only very slowly, so when finally the pain appears, the cryogenic burn is already accomplished. Cold burns are malicious, as owing to an anesthetic effect of the coldness, pain appears only after a while, when the burn has already taken place.

In spite of a broad dissemination in TV and the Internet, this practice remains dangerous for two main reasons: bringing larger parts of the human body in contact with cryogenic liquid leads to accelerated vaporization of the cryogenic liquid, the consequence of which is a highly possible oxygen deficiency of the environment, and last but not least, due to contamination of the wetted skin, the generation of the isolating layer is not always the same, with possibly spots that remain with less or without isolation of the vapor layer, leading to severe burns of these parts.

Working with LIN should always include the use of oxygen monitors, safely indicating any oxygen deficiency to avoid asphyxiation.

So every operation that supports vigorous vaporization of nitrogen can be extremely dangerous but it can be detected by the appearance of fog, generated by contact of the cold nitrogen with moist ambient air. The fog itself is already enriched to a certain degree with nitrogen, so the contact might be harmful, too.

Liquid oxygen has the same properties as cryogenic nitrogen in terms of the effect of the cryogenic temperature. The contamination of the vicinity of a liquid oxygen storage container after spillage of liquid oxygen shows similar serious implications though by oxygen enrichment. Oxygen is a violent accelerator of fire and burns, so an enrichment of oxygen dramatically changes well-known reactions or well-known habits of all materials in a very drastic way.

The term *fire enhancer* does not give a real impression of what is happening in oxygen-enriched atmospheres: materials, which under ordinary conditions, do *not* burn or are regarded as being inert, totally change their behavior.

Things become difficult if the enrichment takes place over a period of time, for example, by a small leak at a container for liquid oxygen stored in a confined space, open oxygen valves of MGPSs or similar occurrences. Before the danger is noticed, every fabric in the room is saturated with oxygen and it would take time to clean clothes from oxygen. Before that, any flame would ignite that mixture, burning down in milliseconds because of the oxygen hidden in the tissue.

7.5 Gases Tapped from Piping Systems: Cross-Contamination and Contamination by Insufficient Handling, Memory Effects

In today's hospitals, the supply of oxygen and air to the patient is usually managed with a central gas supply system (MGPS, Figure 7.8). Liquid medicinal oxygen is vaporized outside of the building and led with a constant pressure into the piping net of the hospital, where it is branched in the different levels and rooms, while

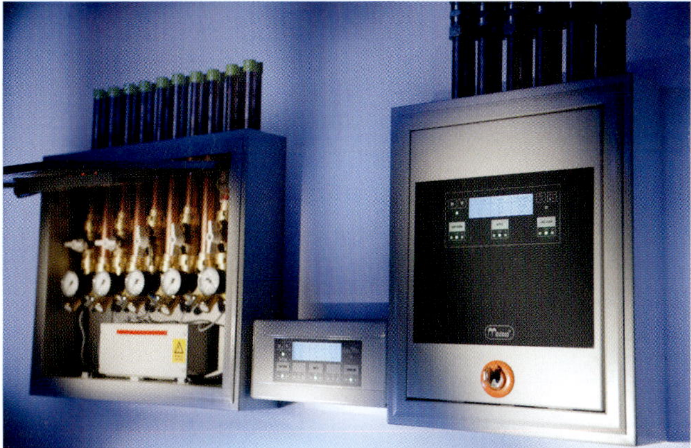

Figure 7.8 Control box for an MGPS (example for level box).

compressed medicinal air is produced at the site, purified, and fed into the piping system.

Actually this means that a drug is transported via the piping system to be inhaled by the patient. Under the present legislation in the European Union, this case applies to the regulation about medical products.

The principle may differ in detail; the great frame being defined by EN ISO 7396-1 [53]. Here the necessary precautions and responsibilities are described to operate such a huge medical device.

It has to be taken into account that the MGPS is a medical product following the MDD, the medical device directive (93/42/EEC, consolidated 2007/47/EC (47)): It is part of the operator's obligations to restrict access, service, and maintenance of the MGPS to qualified people and to follow the operating instructions according the EN-ISO standard.

Following the EN ISO 7396-1, any MGPS is to be designed as a medical product, all parts carrying the EC sign or having a comparable, checked, and proven quality. By application of a risk management procedure possible specific risks are identified in advance, evaluated and assessed, and mitigated by the right measures. The most important risk points may be the following (examples): used hardware: are materials and elements of the MGPS conforming to standard, carrying epsilon-signs or equivalent? To achieve a reliable basis an appropriate documentation during the mounting of the MGPS has to be set up before beginning of the assembly.

After the installation of a new MGPS extended testing of the pipes along the guideline of the EN ISO 7396-1 is mandatory to be sure that the requirements of the standard are completely realized in the MGPS: every wall socket has to be tested for the right gas and the required purity. To achieve reliable results an appropriate documentation during the mounting of the MGPS has to be established before beginning of the assembly.

Next risk point is a necessary opening of the piping for service or maintenance purposes, always connected with a shut down of the wall sockets in that vicinity.

Any service or maintenance thus has to be announced to the responsible person, any opening of the system has to be documented and followed by a formal release, securing the quality of the gas arriving at the patient. Changes in part of the device have to be checked and agreed upon in a change management procedure.

References

1. (a) Riedel, E., and Schram, J. (1990) *Gas aktuell*, **40**, p. 5; (b) Szabadnery, F. (1966) *Geschichte der analytischen Chemie*, Vieweg, Braunschweig, p. 5.
2. Walden, P. (1952) *Chronologische Übersichtstabellen*, Springer, Berlin.
3. Cavendish, H. (1783) *Phil. Trans.*, **73**, 106.
4. Süssmann, O. (2006) portAL,1, p. 20 and others (10,14)
5. Comment by the author based on (9)
6. von Linde, C. (1896) Process and apparatus for liquefying gases or gaseous mixtures, and for producing cold, more particularly applicable for separating oxygen from atmospheric air. GB189512528 (A), May 16 1896 (among others).
7. Claude, G. (1906) Apparatus for the liquefaction of air. US Patent 881176 A, Feb. 23 1906 (among others).
8. Nipper, N. and Hiller, H. (2003) *Gas aktuell*, **65**, pp. 8–13.
9. (a) Sittkus, A., Stockburger, H. (1976) *Naturwissenschaften* **63**, 266–272; (b) Haberer, K. (1986) *Umweltradioaktivität und Trinkwasserversorgung*, **gwf 127**, (12), p. 597–603; (c) Cauwels, P., Buysse, J., Poffijn, A., and Eggermont, G. (2001) Study of the atmospheric 85Kr concentration growth in Gent between 1979 and 1999. *Radiation Physics and Chemistry*, **61**, S. 649–651. doi: 10.1016/S0969-806X(01)00361-9.
10. Zellner, R. (ed.) (2011) *Chemie über den Wolken*, Wiley-VCH, Weinheim.
11. Neumüller, O.-A. (ed.) Römpps Chemie Lexikon, (1983) Bd **3**, Half-life of Krypton-85: 10,76a.
12. Hollemann, A.F. and Wiberg, E. (1964) *Lehrbuch der anorganischen Chemie*, Walter de Gruyter & Co, Berlin.
13. Hiller, H. (2002) *Focus on Gas*, **18**, p. 7–10.
14. Riedel, E. (ed.) (2008) *1x1 der Gase*, Air Liquide Deutschland GmbH.
15. Klebe, U., Thiedecke, L., Stölcker, T., (2002) *Focus on Gas*, **19**, p. 4–9.
16. (a) Ehrensberger, C. (2013) *Nachrichten aus der Chemie*, **61**, 110–112; (b) Ullmanns Encyklopädie d. techn. Chemie (1975) 4. Auflage, Verlag Chemie, Weinheim.
17. Directive 2001/82/EC of the European Parliament and of the Council on the Community code relating to veterinary medicinal products. *Official Journal*, **L311**, 1.
18. Directive 2001/83/EC of the European Parliament and of the Council on the Community code relating to medicinal products for human use. *Official Journal*, **L311**, 67.
19. Council Directive 93/42/EEC concerning medical devices. *Official Journal*, **L 169**, 1.
20. European Commission (2003) Chapter 1 Pharmaceutical Quality System, in *EudraLex - The Rules Governing Medicinal Products in the European Union*, Volume 4: EU Guidelines for Good Manufacturing Practice for Medicinal Products for Human and Veterinary Use, European Commission, Brussels.

Medical Gases: Production, Applications and Safety, First Edition. Hartwig Müller.

21. Niklas, H. (2012) *Krankenhauspharmazie*, **33**, 326–330 and loc. cit. MEDDEV 2, 1/3rev.3, March 2010.
22. Ullmann, F. (1975) *Enzyklopädie der technischen Chemie*, 4. Auflage, Verlag Chemie, Weinheim.
23. Ewald, R., Schreckenberg, W. (1983) *Gas aktuell*, **26**, p. 29.
24. Müller, R. (1997) *Gas aktuell*, **53**, p. 15–22.
25. Neumüller,, O.-A. (ed.) (1981) *Römpps Chemie-Lexikon*, Franckhsche Verlagshandlung, Stuttgart, Bd. **2**.
26. Neumüller, O.-A. (ed.) (1983) *Römpps Chemie Lexikon*, Franckhsche Verlagshandlung, Stuttgart, Bd. **3**.
27. EIGA (2011) Safety Information 24/11/E: Carbon Dioxide Physiological Hazards. CO2 is directly influencing the breathing centre in the brain. The higher the CO2-concentration (up to a few thousand ppm) the higher the breathing frequency will rise.
28. Davy, H. (1800) *Researches, Chemical and Philosophical; Chiefly Concerning Nitrous Oxide, or Dephlogisticated Nitrous Air, and its Respiration*, J. Johnson, London.
29. Grefer, J. (1982) *Gas aktuell*, **23**, p. 23–27.
30. Downie, A. (1997) *Industrial Gases*, Blackie Academic & Professional, London, p. 509ff, 506.
31. Downie, N.A. (1997) *Industrial Gases*, Blackie Academic & Professional, London, p. 509ff.
32. Krinninger, K.-D. (1996, 2001) *Kohlendioxid-Kohlensäure*, Verlag Moderne Industrie, Landsberg am Lech.
33. Grabhorn, G., Reimann, B., Müller, H. (2001) Focus on Gas, **16**, p. 31–36.
34. Zschunke, A. (ed.) (2000) *Reference Materials in Analytical Chemistry*, Springer, p. 203.
35. (2009) *Die Geschichte der inhalativen Sauerstofftherapie in Deutschland*, Christina Kossobutzki, Lübeck.
36. Müller, H. (2004) EIGA 2004, Strasbourg, 300 bar: Challenges and Solutions for Safety within the Gases Industry.
37. Downie, N.A. (1997) *Industrial Gases*, Blackie Academic & Professional, London, p. 237.
38. (2010) Council Directive 1999/37/EC on transportable pressure equipment. Official Journal L138/20, 1.06.1999 repealed by Directive 2010/35/EU of the European Parliament and of the Council on transportable pressure equipment. *Official Journal*, **L165**, 1.
39. (2003) Directive 97/23/EC of the European Parliament and of the Council on the approximation of laws concerning pressure equipment. Official Journal L181, 9.07.1997, 1; amended by regulation (EC) No 1882/2003. *Official Journal*, **L284**, 1.
40. (2009) Directive 2009/105/EC of the European Parliament and of the Council relating to simple pressure vessels. *Official Journal*, **L 264**, 12.
41. (1975) Directive of the Council on the approximation of the laws of the member states relating to aerosol dispensers. *Official Journal*, **L147**, p. 40.
42. (1999) Verordnung über Druckbehälter, Druckgasbehälter und Füllanlagen (Druckbehälterverordnung) Letzte Fassung 23.06.1999. Bundesgesetzblatt I, S. 1436, expired 1 January 2003.
43. Transport Canada (02-12-2012) UN cylinders, UN Tubes, ZN cryogenic receptacles, and multi-element gas containers (MEG's). www.tc.gc.ca, Safety/containers. Last access: https://www.tc.gc.ca/eng/tdg/moc-menu-1189.html (06-01-2015).
44. Following the recommendation of the manufacturer, Aluminum cylinders should not be heated excessively, long time heating should remain below 100°C, so that no changes in the crystallization patterns can occur. In opposite to steel cylinders, Aluminum cylinders do not undergo changes in tensility or strength down to temperatures until −196°C.
45. Zeitung, F.A. (2000), 21.11.2000.
46. (1965) Council Directive 65/65/EEC on the approximation of provisions laid down by law, regulation or administrative action relating to medicinal products. *Official Journal*, (No 22 of 9. 2.) p. 369.

47. (2001) Directive 2001/82/EC of the European Parliament and of the Council on the Community code relating to veterinary medicinal products. *Official Journal*, **L311**, 1.
48. European Directorate for the Quality of Medicines & Healthcare, (2008) *European Pharamcopoeia 6.0*, p. 423 ff.
49. Grant, W.J. (2005) Medical Gases, Their Properties and Uses, BOC 2nd edition, p. 10.
50. Mueller, H., Dovermann, F. (2012) PortAL, **12**, p. 11–13.
51. Air Liquide – Taema – Integral valve Compact® G2 – 2004 Product Leaflet.
52. Plass, M. (2013) PortAL, Air Liquide Healthcare, **15**, p. 20–21.
53. CEN – European Committee for Standardization (2007) Medical Gas Pipeline Systems — Part 1 – Pipeline systems for compressed medical gases and vacuum (EN ISO 7396-1:2007).
54. (2004) Standard Operating Procedure Air Liquide Deutschland GmbH.
55. a) Air Liquide SOP 2004: Cryosampling with Cosmodyne Sampler, b) Sampler by Cosmodyne LLC, 3010 Old Ranch Parkway, Seal Beach, CA 90740, USA; www.cosmodyne.com.
56. Riedel,E. and Schram, J. (1990) *Gas aktuell*, **40**, p. 6.
57. European Directorate for the Quality of Medicines & Healthcare, (2013) European Pharmacopoeia, 8.1, Chapter 2.
58. European Directorate for the Quality of Medicines & Healthcare, (2013) European Pharmacopoeia, 8.1, Chapter 2.24, p. 38.
59. Müller, H. (1997) Krankenhauspharmazie, **18**, Nr. 6, p. 273–282.
60. European Directorate for the Quality of Medicines & Healthcare, (2013) European Pharmacopoeia, 8.1, Chapter 2.2.8, p.43.
61. Ettre, L.C. (1990) *LC-GC International*, **3**, 28.
62. European Directorate for the Quality of Medicines & Healthcare, (2013) European Pharmacopoeia, 8.1, Chapter 2.5.26.
63. (1968) *Luminescence in Chemistry*, van Nostrand Co Ltd, London.
64. European Directorate for the Quality of Medicines & Healthcare, (2013) European Pharmacopoeia, 8.2, Monograph Oxygen 6.6/0417.
65. European Directorate for the Quality of Medicines & Healthcare, (2013) European Pharmacopoeia, 8.1, Chapter 2.5.28 p. 145.
66. a) Hartmann & Braun AG (ca. 1995). Sauerstoffmessung / K.H. Koch, (1997) *Industrielle Prozeßanalytik*, Springer-Verlag, Heidelberg.
67. Hilscher, W., Riedel, E., Schmicker, D., (1986) *Gas aktuell*, **32**, p. 31.
68. European Directorate for the Quality of Medicines & Healthcare, (2013) European Pharmacopoeia, 8.1, Monograph Air, Medicinal, 01/2009:1238, p. 1331.
69. a) European Directorate for the Quality of Medicines & Healthcare, European Pharmacopoeia, 8.1 Chapter 2.1.6 Test tubes & b) Arzneibuch Kommentar, Bracher, Heisig, Langguth, Mutschler, Rücker, Scriba, Stahl-Biskup, Troschütz (eds) 2013, Wissenschaftliche Verlagsgesellschaft Stuttgart.
70. Draeger Tubes & CMS Handbook (2011) *Soil, Water, and Air Investigations*, 16th edn, Dräger Safety AG & CO, KGaA, Lübeck.
71. a) European Directorate for the Quality of Medicines & Healthcare, (2013) Annual Report; b) EIGA (2011) Comparison of European, US & Japanese Pharmacopoeia Monographs for Medicinal Gases. EIGA MGC Doc 152/08/E.
72. European Directorate for the Quality of Medicines & Healthcare, (2014) *European Pharmacopoeia 8.0*, 01/2014.
73. (a) Priestley, J. (1974) *Das Buch der grossen Chemiker*, Band **1**, (Prof. Georg Lockemann), Verlag Chemie GmbH, Weinheim, p. S. 263 ff; (b) Scheele, C.W. (1943) Ein Gedenkblatt zu seinem 200. Geburtstage. *Zeitschrift für Anorganische und Allgemeine Chemie*, **Jg. 250** (Heft 3-4), S. 230–235, ISSN: 1521-3749.
74. Martin, L. (2011) Oxygen Therapy: The First 150 Years; Curiosities, Quackeries, and Other Historical Trivia; A Chronology from Priestley to Haldane, Based Mainly on Original Sources, Mt.

Sinai Medical Center, Cleveland, OH (internet, last update 2011).

75. Kossobutzki, C. (2009) Die Geschichte der inhalativen Sauerstofftherapie in Deutschland, Lübeck: Cit 2: Leigh, J.M. (1974) Early treatment with oxygen. The Pneumatic Institute and the panacea. *Anaesthesia*, **29**, 194–208.
76. (1994) Decision of the Council and the Commission on the conclusion of the Agreement on the European Economic Area (EEA)...(94/1/CE, ECSC, EC). *Official Journal*, L 1/1 of 3.1.1994.
77. ICH *www.ich.org* (accessed 24 October 2014).
78. European medicines Agency *www.ema.europa.eu* (accessed 24 October 2014).
79. European Commission (2012) Chapter 1 Pharmaceutical quality system, in *EudraLex V29 – The Rules Governing Medicinal Products in the European Union*, Volume 4: EU Guidelines for Good Manufacturing Practice for Medicinal Products for Human and Veterinary Use, European Commission, Brussels.
80. CEN – European Committee for Standardization (2012) EN ISO 13485:2012+AC:2012, Medical Devices – Quality Management Systems – Guidance on the Application.
81. International Conference on Harmonisation of Technical Requirements for Registration of Pharmaceuticals for Human Use (2005) ICH Tripartite Guideline Quality Risk Management Q9 – Step 4 Version.
82. EMA – European Medicines Agency (2014): EMA/CHMP/ICH/214732/2007, ICH guideline Q10 on pharmaceutical quality system – Step 5.
83. European Commission (2013) Personnel, in *EudraLex V29 – The Rules Governing Medicinal Products in the European Union*, Chapter 2 Personnel: Volume 4: EU Guidelines for Good Manufacturing Practice for Medicinal Products for Human and Veterinary Use, European Commission, Brussels.
84. European Commission (2014) Chapter 3: Premises and equipment, in *EudraLex V29 – The Rules Governing Medicinal Products in the European Union*, Volume 4: EU Guidelines for Good Manufacturing Practice for Medicinal Products for Human and Veterinary Use, European Commission, Brussels.
85. European Commission (2010) *EudraLex V29 – The Rules Governing Medicinal Products in the European Union*, Volume 4: EU Guidelines for Good Manufacturing Practice for Medicinal Products for Human and Veterinary Use Annex 6, pt 7, 9 and 45, European Commission, Brussels.
86. European Commission (2010) Chapter 4: Documentation, in *EudraLex V29 – The Rules Governing Medicinal Products in the European Union*, Volume 4: EU Guidelines for Good Manufacturing Practice for Medicinal Products for Human and Veterinary Use, European Commission, Brussels.
87. European Commission (2013) Chapter 5: Production, in *EudraLex V29 – The Rules Governing Medicinal Products in the European Union*, Volume 4: EU Guidelines for Good Manufacturing Practice for Medicinal Products for Human and Veterinary Use, European Commission, Brussels.
88. European Commission (2014) Chapter 6: Quality control, in *EudraLex V29 – The Rules Governing Medicinal Products in the European Union*, Volume 4 EU Guidelines for Good Manufacturing Practice for Medicinal Products for Human and Veterinary Use Part 1, European Commission, Brussels.
89. European Commission (2012) Chapter 7: Outsourced activities, in *EudraLex V29 – The Rules Governing Medicinal Products in the European Union*, Volume 4, EU Guidelines for Good Manufacturing Practice for Medicinal Products for Human and Veterinary Use, European Commission, Brussels.
90. European Commission (2005) Chapter 8 Complaints and product recall, in *EudraLex V29 – The Rules Governing Medicinal Products in the European Union*, Volume 4 EU Guidelines for Good Manufacturing Practice Medicinal Products for Human and Veterinary Use, European Commission, Brussels.

91. European Commission (2010) Chapter 9 Self Inspection *EudraLex V29 – The Rules Governing Medicinal Products in the European Union*, Volume 4 EU Guidelines for Good Manufacturing Practice Medicinal Products for Human and Veterinary Use, European Commission, Brussels.
92. European Commission (2005) *EudraLex V29 – The Rules Governing Medicinal Products in the European Union*, Volume 4: Good Manufacturing Practice Medicinal Products for Human and Veterinary Use, Part II: Basic Requirements for Active Substances used as Starting Materials, European Commission, Brussels.
93. European Commission (2010) *EudraLex V29 – The Rules Governing Medicinal Products in the European Union*, Volume 4: Good Manufacturing Practice Medicinal Products for Human and Veterinary Use, Part II: Basic Requirements for Active Substances used as Starting Materials, Chapter 7, European Commission, Brussels, p. 20.
94. European Commission (2010) *EudraLex V29 – The Rules Governing Medicinal Products in the European Union*, Volume 4: Good Manufacturing Practice Medicinal Products for Human and Veterinary Use, Part II: Basic Requirements for Active Substances used as Starting Materials, Chapter 11.5, European Commission, Brussels, p. 28.
95. EMA-European Medicines Agency – Committee for Human Medicinal Products (CHMP) (2007) Guideline on Medicinal Gases: Pharmaceutical Documentation, Note for Guidance CPMP/QWP/1719/00 Rev. 1.
96. European Commission (2010) *EudraLex V29 – The Rules Governing Medicinal Products in the European Union*, Volume 4: Good Manufacturing Practice Medicinal Products for Human and Veterinary Use Part II: Basic Requirements for Active Substances used as Starting Materials, Chapter 11.7, European Commission, Brussels.
97. European Commission (2010) *EudraLex V29 – The Rules Governing Medicinal Products in the European Union*, Volume 4: Good Manufacturing Practice Medicinal Products for Human and Veterinary Use Explanatory Notes on the preparation of a Site Master File, European Commission, Brussels.
98. ICH (2009) Harmonized Tripartite Guideline Pharmaceutical Development Q8(R2) Current Step 4 Version, dated August 2009.
99. (2013) Commission Guideline of 7 March 2013 on Good Distribution Practice of medicinal products for human use. *Official Journal*, **C 68**, 1.
100. (1994) Commission Guideline on Good Distribution Practice of medicinal products for human use. *Official Journal*, **C 63**, 4.
101. (2013) Commission Guideline of 5 November 2013 on Good Distribution Practice of medicinal products for human use. *Official Journal*, **C 343/4**, 1.
102. (2011) Directive 2011/62/EU of the European Parliament and of the Council amending Directive 2001/83/EC on the Community code relating to medicinal products for human use, regards the prevention of the entry into the legal supply chain of falsified medicinal products. *Official Journal*, **L 174**, 74.
103. a) Regulation (EC) No 726/2004 of the European Parliament and of the Council of 31 March 2004; b) Regulation (EU) No 520/2012 of the Commission of 19 June 2012 on the performance of pharmacovigilance activities.
104. (2001) Directive 2001/83/EC of the European Parliament and of the Council on the Community code relating to medicinal products for human use. *Official Journal*, **L311**, 67 Article 1(17).
105. (2001) Directive 2001/83/EC of the European Parliament and of the Council on the Community code relating to medicinal products for human use. *Official Journal*, **L311**, 67. Article 80(h).
106. (2001) Directive 2001/83/EC of the European Parliament and of the Council on the Community code relating to medicinal products for human use. *Official Journal*, **L311**, 67. Article 79(b).
107. (2001) Directive 2001/83/EC of the European Parliament and of the Council on the Community code relating

to medicinal products for human use. *Official Journal*, **L311**, 67. Article 79(a).

108. (2001) Directive 2001/83/EC of the European Parliament and of the Council on the Community code relating to medicinal products for human use. *Official Journal*, **L311**, 67. Article 1(17a).
109. CEN – European Committee for Standardization (2013) EN ISO 14971:2012 Medical Devices – Application of Risk Management to medical devices.
110. Guideline (2012) of the Commission on good pharmacovigilance practices (GVP) – Modules I-XVI to cover major pharmacovigilance processes, EMA/204715/2012 (Rev 1).
111. EIGA (2008) Asphyxiation Risk in Open Cryo Cabins with Cooling by Means of Direct LIN Injection – EIGA Safety Information 19/2008/E.
112. EIGA (2003) Dangers of Asphyxiation Leaflet.
113. EIGA (2003) Presentation PR 01/2003 PDF.
114. (a) (2012) REGULATION (EU) No 1027/2012 OF THE EUROPEAN PARLIAMENT AND OF THE COUNCIL of 25 October 2012, amending Regulation (EC) No 726/2004 as regards pharmacovigilance; (b) (2012) DIRECTIVE 2012/26/EU OF THE EUROPEAN PARLIAMENT AND OF THE COUNCIL of 25 October 2012, amending Directive 2001/83/EC as regards pharmacovigilance.
115. Steel cylinders are provided by e.g. www.tenaris.com, www.faber-italy.com, www.eurocylinders.com, www.worthingtoncylinders.com
116. Aluminum cylinders for high pressure gases e.g. www.luxfercylinders.com
117. (1967) Council Directive 67/548/EEC on the approximation of laws, regulations and administrative provisions relating to the classification, packaging and labelling of dangerous substances. *Official Journal*, **P 196**, 1.
118. (1999) Directive 1999/45/EC of the European Parliament and of the Council concerning the approximation of laws, regulations and administrative provisions relating to the classification, packagingand labelling of dangerous preparations. *Official Journal of European Community*, **L200**, 1–68.
119. (2008) REGULATION (EC) No 1272/2008 OF THE EUROPEAN PARLIAMENT AND OF THE COUNCIL of 16 December 2008 on classification, labelling and packaging of substances and mixtures, amending and repealing Directives 67/548/EEC and 1999/45/EC, and amending Regulation (EC) No 1907/2006.
120. EIGA (2014) IGC Doc 133 Cryogenic Vaporisation Systems, Prevention of Brittle Fracture of Equipment and Piping. IGC Doc 133/14/E.
121. (2008) EIGA Safety Information 19/08/E Asphyxia – Effects and Symptoms of reduced O_2 concentration (Vol%).
122. REGULATION (EC) No 1907/2006 OF THE EUROPEAN PARLIAMENT AND OF THE COUNCIL of 18 December 2006 concerning the Registration, Evaluation, Authorisation and Restriction of Chemicals (REACH).

Abbreviations

AA	Aluminum alloys
ACAA	Agreement on Conformity Assessment and Acceptance of Industrial Products
AFNOR	Association francaise de normalisation
APIs	Active pharmaceutical ingredients
ASU	Air separation unit
BAM	Bundesanstalt für Materialprüfung
BSI	British standard institute
CAPA	Corrective Action/Preventive Action
CLP-regulation	Classification, Labeling and Packaging-regulation
DIN	Deutsches institut für normung
EC	European Community
EEC	European Economic Community
EU	European Union
EIGA	European Industrial Gas Association
EMA	European Medicines Agency
E.P.	European Pharmacopoeia
FID	Flame ionization detector
GC methods	Gas chromatorgraphic methods
GDP	Good distribution practice
GMP-rules	Good manufacturing practice-rules
GVP	Good pharmacovigilance practices guidelines
HoP	Head of Production Department
HoQ	Head of Quality Control (HoQ)
ICH	International Conference on Harmonization
IGV	Dt. Industriegaseverband
IR	Spectrometryinfrared spectrometry
LIN	Liquid nitrogen
MAH	market authorization holders
MAs	market authorizations
MDD	medical device directive
MGPS	medicinal gas pipeline systems

Medical Gases: Production, Applications and Safety, First Edition. Hartwig Müller.

MOC	management of change-process
MRAs	Mutual Recognition Agreements
MRSA	Methicillin-resistant *Staphylococcus aureus*
MRT	Magnetic resonance tomography
NDIRs	Nondispersive infrared instruments
NRV/PRV	Valves with Integrated Residual Pressure/Nonreturn Cartridge
Ph. Eur.	European Pharmacopoeia
PIC/S	Pharmaceutical Inspection Cooperation/Scheme
PQR	Product Quality Review
PQS	Pharmaceutical Quality System
PSA	Personal safety accessories
QC	Quality control
QMSs	Quality management systems
QRM	Quality risk management
REACH	Registration, Evaluation, Authorisation and Restriction of Chemicals
SDS	Safety data sheet
SMF	Site master file
ss	Single sampling
THC	Thermal heat conductivity
TPED	Transportable pressure equipment directive

Index

Medical Gases: Production, Applications and Safety, First Edition. Hartwig Müller.